WATKINS
REYNOLDS
MATTHEWS

WATKINS REYNOLDS MATTHEWS

Biography of a Texas Rancher

By Lawrence Clayton

Drawings by Paul Cameron Smith

EAKIN PRESS ★ Austin, Texas

FIRST PRINTING 1990
SECOND EDITION

Published in the United States of America
By Eakin Press
A Division of Sunbelt Media, Inc.

ISBN 0-89015-950-5

10 9 8 7 6 5 4 3 2 1

Library of Congress Cataloging-in-Publication Data

Clayton, Lawrence, 1938–
Watkins Reynolds Matthews : a biography / by Lawrence Clayton : drawings by Paul Cameron Smith.
p. cm.
Includes bibliographical references and index.
ISBN 0-89015-950-5 : $16.95
1. Matthews, Watkins Reynolds, 1904–. 2. Ranchers – Texas – Brazos River Region – Biography. 3. Ranch life – Texas – Brazos River Region. 4. Brazos River Region (Tex.) – Biography. I. Title.
F392.B842M364 1994
976.4'06'092 – dc20
[B] 93-45374
CIP

CONTENTS

ILLUSTRATIONS

ACKNOWLEDGMENTS

I appreciate the help of four secretaries who typed the manuscript: Dorothy Kiser, Carol Donald, Joy Cron, and Charlene Archer. Great appreciation is extended to my longtime friend and mentor, Dr. Lloyd D. Huff, who graciously read the manuscript and, as usual, kept me from committing publicly several errors. Those that have crept in are my own. I also wish to acknowledge the suggestion and support of William Curtis of the Hardin-Simmons University Business faculty and Randy Armstrong of the Mass Communication faculty for encouraging me to complete this manuscript. The drawings by Paul Smith are certainly a welcome addition to the project. For the second edition, I owe thanks to Sonja Irwin Clayton for the use of her photographs and to Janet Neugebauer of the Southwest Collection of Texas Tech University for aid in securing photographs from the Matthews family archives.

I want to express appreciation for the help and support of my wife, Sonja, and two daughters, DeLys Clayton Mitchell and Lea Clayton, as well as to my parents, Mr. and Mrs. Donald H. Clayton, without whose support I would not have been in a position to undertake the work at all.

My final note of thanks goes to Watt Matthews, without whose assistance through the interviews I could not have completed the volume. His gracious hospitality made this project pleasant and productive as well as memorable.

A very young Watt Matthews, in 1905, taking advantage of a chance to try his luck at fishing.

Watt as a young man in front of the big red barn at the headquarters.

AUTHOR'S PREFACE TO THE SECOND EDITION

The second edition of this work is not simply a reprinted version of the original text. Instead, the type has been reset so that new information about the life of Watkins Reynolds Matthews could be woven into the existing text. This step was necessary because, despite his age, in the more than four years since I completed the original version, Watt has not been idle. His efforts with the *Fort Griffin Fandangle*, historical preservation, and ranching as well as his very active social life have continued unabated. The scope of these activities warrants an updating of the life story of this remarkable man.

The first edition of just over five hundred copies including four hundred and fifty of the standard edition, fifty of the deluxe edition, ten of the numbered gourd edition, and three in special bindings sold out in less than a year. The few copies that have appeared on the rare book market have sold for almost twice the original price. This continuing interest in the book by Watt's many admirers and his prominence as a Texan are certainly reasons enough to make the work available in an edition more affordable than was possible with the limited number of copies in the first edition.

My admiration of and respect for Watt have continued to grow, and the enthusiastic reception of the volume has been gratifying. It is indeed a pleasure to bring an updated version of the story to an appreciative audience.

I am grateful to Ed Eakin for his superb effort in this edition and to those who encouraged me to update the volume.

Watt in a festive mood as a young man.

AUTHOR'S PREFACE TO THE FIRST EDITION

A biography of Watkins Reynolds Matthews is a well-deserved tribute to one of the truly unique citizens of West Texas. I had known of Mr. Matthews since the late 1950s and had long believed his life and accomplishments worthy of a book-length study. I did not really get to know him personally until in the mid-1980s. At that time I began studying cowboy and ranch life along the Clear Fork of the Brazos River in Shackelford, Throckmorton, and Jones counties of West Texas. I soon learned that *Watt Matthews, ranching,* and *Clear Fork* are virtual synonyms, for Watt has managed a solid ranching operation and played an active role in the development of the Clear Fork country for many years, as did his family before him.

My first interview with this modern-day frontiersman was on Memorial Day of 1984, when I went to Lambshead Ranch to observe and photograph him and his cowboys working cattle. After the working for that day ended, he and I sat in the cookhouse, where he responded to some general questions about ranching in the area where he has run this large ranch since the 1940s. In 1985 I began seriously researching his life to prepare an entry I was asked to write for the revised edition of *The Handbook of Texas*. The more I learned, the more impressed I became with the quiet sincerity and calm demeanor which veiled an alert, active man already in his eighties. Later I talked with, interviewed, and worked alongside this veteran cowman and his cowboys on several occasions and recall with great fondness the experience and the memories.

At the 1986 meeting of the West Texas Historical Association in Sweetwater, I read a paper on Watt's remarkable accomplishments in historical preservation. This presentation included numerous slides made by Sonja Irwin Clayton of the structures on the ranch. Since Watt could not attend, John Burns, a nephew with long-standing interest in the ranch, came to see the presentation at the Sweetwater meeting. An article incorporating this material later appeared in the Winter, 1986, issue of *Heritage Magazine*, the organ of the Texas State Historical Foundation. Watt himself was kind enough to be complimentary of the article. On January 22, 1987, I went again to Lambshead with Dr. B. W. Aston, professor of history and director of the Richardson Research Center at Hardin-Simmons University in Abilene. There we videotaped an interview with Watt in the cookhouse at the headquarters and stayed on for a time of casual conversation and a steak dinner prepared by John White, one of the longtime cooks at the ranch.

I returned on February 6, 1987, with Robert Reinhold, bureau chief of the Houston office of the *New York Times*. Reinhold was so intrigued with Watt's personality, philosophy, and lifestyle that he rearranged plans and took a later flight back to his office in Houston. The positive response of one so cosmopolitan in outlook and background affirmed that my interest in Watt was well directed. Reinhold's visit later resulted in three articles complete with pictures which appeared in the April 1, 1987, issue of the *Times*. Other trips, especially the one on January 20, 1988, where two young artists visited with and photographed Watt, proved the genuine warmth of Watt's hospitality. The drawing of Watt by Paul Smith, used as the frontispiece of the volume, is the result of this enjoyable afternoon.

The materials with which I was able to work were rich in scope and content. I learned from Dr. David Murrah, director of the Southwest Collection at Texas Tech University in Lubbock, that the Matthews family had a number of scrapbooks and photograph albums that spanned much of Watt's life. In order to view them, I made another visit to the ranch on March 28, 1987, where I perused with amazement collections that covered many events significant to the area and the ranch. In addition, I used the occasion to check wording and dates on various documents. Watt generously welcomed the visit, even though he had to be away from the ranch at the time to attend the funeral of an old friend.

My appreciation of Watt has continued to grow throughout my research. I had earlier chosen him as the lead figure in *Clear Fork Cowboys*, for there was never any doubt about who should begin that collection detailing the life and work of area cowboys and cowmen along the winding, reddish river. The subject of numerous other articles, sketches, photographs, and references, and the recipient of numerous awards and honors, Watt is obviously a giant in his time, a man who has left deep tracks along and beyond the Clear Fork of the Brazos ranch country he calls home. A statement in an article on Watt in the newsletter of the Center of Texas Studies at the University of North Texas State in Denton suggests that Watt "should be declared a national treasure,"[1] a statement I heartily endorse. Here, then, is a modern frontiersman with a fascinating story, which I appreciate the opportunity to tell.

The serious pose of a well-dressed ranch patriarch.

The aging patriarch in his characteristic work hat. Note the gourd brand of Reynolds Brothers Cattle Company on his shirt pocket.

WATKINS
REYNOLDS
MATTHEWS

CHAPTER I

THE BACKGROUND OF WATT MATTHEWS

The development of the ranching culture along the Clear Fork of the Brazos River in Shackelford and Throckmorton counties of West Texas is recorded in an outstanding family saga entitled *Interwoven: A Pioneer Chronicle.* The work of Sallie Reynolds Matthews, Watt's remarkable mother, the book tells the story of the Reynolds and Matthews families and the details and personalities that helped shape the economic and social development of this region. The last sentence of this narrative says, in part, "So the closing year of the century was marked for us by the birth of our last child."[1] He was born in nearby Albany, not on the ranch he was to come to love. The family named this child Watkins Reynolds Matthews, thus marking him with names important to this frontier family: *Watkins* was a familiar name on the Reynolds side of the family, and the middle name, *Reynolds*, was, of course, the maiden name of the boy's mother. Watt, as he came to be called by his friends—which amounts to almost everyone who meets him, for virtually all who know him feel a closeness to him—was to make an important contribution to life in the Southwest. Existence there has been altered for the better by the efforts of this remarkable man whose birth is mentioned in this almost incidental remark at the close of his mother's book on her life and family and the development of a bonafide cattle culture in West Texas.

Privileged indeed are those who are born into a life that fits them as well as West Texas ranch life has suited Watt Matthews. Through actively pursuing his varied interests, he has become widely known as a rancher, oilman, philanthropist, historical preservationist, and cattle country gentleman. A Princeton University graduate, Watt has provided dynamic leadership and support in several areas of life associated with his West Texas ranching interests while managing the day-to-day operations of Lambshead Ranch.

In order to appreciate Watt's accomplishments, especially those resulting from his orientation toward history, one needs to know the context in which he was born and grew up. The story is indeed an intriguing one, as a brief sketch demonstrates.

Watt's family was among the earliest Anglo settlers in the Clear Fork country. The Reynolds and Matthews families, of English, Scotch, and Welsh extraction, followed a familiar pattern of Anglo settlement in Texas. Migrating from the southern states of Alabama and Georgia to Louisiana and then on into East Texas, the two families eventually settled in the Clear Fork area, the Matthews family by 1858 and the Reynolds family by 1859. They were true pioneers, for the area was then sparsely settled by Anglos and constantly threatened by hostile Indians. The Reynolds family had two daughters and several sons. The Matthews family had one son (John A. Matthews, later the father of Watt), and several daughters; hence, the opportunity for these children to intermarry was strong, and before long the two families were interwoven by several such unions. Thus were entwined the lives and fortunes of the two families as they worked alongside others in the region—Gentry, Irwin, Newcomb, Pate, Lynch, Grier, and others.

Members of the group "forted up" in the area during the dangerous later years of the Civil War—the Reynolds family at Fort Davis in Stephens County and the Matthews family at Fort Owl Head. These were not military posts, for all government soldiers had been withdrawn from the Texas frontier to face each other on distant southern battlefields during the War between the States. These West Texas families, isolated on the unspoiled grasslands, refused to leave their homes and growing herds of wild longhorn cattle. For protection, they chose to live in individual picket-style cabins clustered together in a rough square for mutual protection; hence, the term "forting up." These crude but serviceable dwellings were made by driving the sharpened end of small logs upright into the earth to form the walls. No timbers large enough for traditional log cabin construction were available. The stone house that anchored one corner of Fort Davis still stands as mute testimony of these challenging days in frontier Texas. Owned by A. V. Jones, Jr., and Harold Law of Albany, this structure is now outfitted as a hunting lodge.

Even during these dangerous early times, however, the cowmen braved the elements and hostile Indians to keep up their "cow hunts" for the wild cattle that roamed the river bottoms and rocky uplands. As they found unbranded cattle, the cowboys added the animals to the growing herds that roamed at will on the vast, unfenced, and almost treeless prairie where, in good years, the grass rose stirrup deep. After the war had ended and soldiers returned to establish Fort Griffin to protect the settlers, the families moved back onto the grazing land to establish permanent ranches and farm the rich soil along the river. These hardy settlers sank their roots deep into the Clear Fork country.

Sallie Reynolds Matthews notes that the Reynolds family settled briefly at the Stone Ranch near the Clear Fork in 1866–67 while the crews herded cattle on the yet unfenced land. Her brother, Phin, recounted for J. R. Webb, a local historian who made an important contribution by recording the recollections of pioneer settlers, that some people named Knox and Gardner who lived at the house during the Civil War had left some cattle roaming the surrounding range. The Reynolds family bought the cattle from the owners, but a problem emerged. Phin points out, "We had to pay for them twice, for we paid each one of them [Knox and Gardner] for the same 100 head."[2] Later members of the Reynolds family established ranches in other areas of the West as well. Among them are the Long X Ranch at Kent; the Davis Mountain Ranch at Kent; the George Reynolds Ranch at Pecos City; the Glen Reynolds Ranch near Globe, Arizona; the Park Springs Ranch in New Mexico; and other ranches in Colorado, Arizona, and other parts of the West.

Watt's parents, John A. Matthews and Sallie Reynolds, married on Christmas Day in 1876 in what came to be called the

Reynolds house in Reynolds Bend along the Clear Fork of the Brazos. They established a ranch on the north side of the Clear Fork at the head of California Creek, so named because the California Trail that funneled miners to the Gold Rush in the 1840s and 50s ran beside this modest stream. Here in a small stone house began a legacy that continues to be influential in the region. From this ranch, the cow outfit—at that time still largely an almost constantly roving collection of wagons, cowboys, and a remuda of saddle horses—moved eventually to a spot on the south side of the Clear Fork, the present-day site of Lambshead Ranch.

The Matthews family owned no land in the area, nor did anyone else until the late 1870s, when private ownership became the pattern. J. A. Matthews knew the value of land but by 1880 he was, as Watt puts it, "land poor," because of the thousands of acres of land he controlled. He needed operating capital, and to generate it he took in partners in his enterprise for a period of five years. This partnership, which included members of the Reynolds family, used the MO brand, which had already been registered under the name of J. A. Matthews. When the five years ended, the partners split up the land. J. A. Matthews moved to the eastern part of the property to Lambshead and came to use the AV brand, which he registered in 1885.

The other property continued to be part of the West Texas ranching business. Some of the ranch was bought by T. G. Hendrick in the 1940s. Part of it for a time provided a home for several boys who were residents of the Hendrick Home for Children, an organization established by Hendrick. The remainder of the property is a working ranch controlled by terms of a trust with the Texas-American Bank in Fort Worth, with Dub Sims as the bank's representative. Another part of the property, known as the X Ranch at the Reynolds Round Mountain Headquarters along the Clear Fork, became the property of Ross Sloan, a rancher with large holdings in the area and in Stephens County. Now his grandsons Carl and Bill Everett operate the portion of the ranch they inherited; their cousins operate the other portion. Bill Davis bought the partnership property along Paint Creek north of the Clear Fork. The Nail Ranch also covers part of the old MO range. Cattle still roam these rich grasslands as they did in frontier days.

Lambshead Ranch has become by far the best known of these early spreads and now includes some forty thousand acres of rolling ranch land cut by several miles of the welcome waters of the Clear Fork. Named for its founder, Thomas Lambshead, originally from Devon, England, the ranch was established about 1847, just two years after Capt. Randolph Marcy laid out the California Trail through this area. Lambshead was a farmer in a land that has rarely been kind to those who till the soil; droughty years far outnumber those in which rainfall is adequate for farming. He had a hedge against financial disaster, because he also managed the Butterfield Overland Mail Station, established on a creek that bears his name. At the mouth of this creek, a federal Indian agent named Jesse Stem had earlier plowed the first land to be cultivated in Northwest Texas. He was later killed by Kickapoo Indians, and his grieving family returned to Ohio. Not far away George Reynolds was severely wounded in an Indian fight by an arrow which lodged near his spine. Si Hough, a close friend and associate, killed the Indian and took the fancy silver-decorated bridle from the Indian's horse. It was years later that famous trail driver Shanghai Pierce stood beside Mr. Reynolds in St. Louis when a surgeon made an incision in search of the arrowhead. On that occasion, Pierce admonished the surgeon that his repeated efforts to locate the arrowhead would cut the man "to the hollow,"[3] a remark which caused the surgeon to cease his probings. It was not until Mr. Reynolds sat up that the arrowhead became visible in the incision. That steel point, which Mr. Reynolds carried in his body for years, and the bridle taken from the Indian's horse by Hough are presently on display in the National Cowboy Hall of Fame in Oklahoma City.

Across the Clear Fork from Lambshead is the site of Camp Cooper. The outbreak of the Civil War, fought far from the Clear Fork area, almost came here in February of 1861, but Union forces yielded to pressure from state troops under Col. W. C. Dalrymple. The withdrawal of Company I of the First United States Infantry, the last unit at the camp, averted open hostility, which broke out in April at Fort Sumter, South Carolina. The post was the Texas home of Col. Robert E. Lee from 1856–59. Lee called the area a "desert of dullness" in personal correspondence back to his family living in the lush, humid areas of the South to which he compared the semi-arid Texas landscape. Lee went on to become commander of the Department of Texas[4] and later, of course, served as commander-in-chief of the forces of the Confederacy before signing the treaty with Ulysses S. Grant to end a tumultuous period of open warfare that split the United States. Part of his toughening for what lay ahead of him came from his experience in Texas.

This national conflict also caused problems in Texas because hostile Indians assumed that they were winning the battle and stepped up their efforts to drive out the white intruders. A Comanche Indian reservation, abandoned in 1859 under pressure by local citizens led by former Indian Agent John R. Baylor, was also located near what was to become the Matthews property. The temporary relocation of Camp Cooper was to a spot on what became part of Lambshead Ranch, but the garrison returned to the original site before the fort was abandoned in 1861. Only the ruins of the small bake oven remain to mark the spot. Other nearby forts of this pre-Civil War period were Fort Belknap in present-day Young County; Fort Phantom Hill, north of present-day Abilene; Fort Chadbourne, located between the present-day cities of Abilene and San Angelo; and, to the northwest, Fort Preston, located near the Red River to supply the other forts. Since hostile Indians were still a threat, the period after the war saw the founding of Fort Griffin[5] down river from the Camp Cooper site, and Fort Concho in present-day San Angelo. Fort Belknap near present day Newcastle, manned by state troops during the war, was reactivated in April of 1867. Fort Phantom Hill, burned and abandoned, and Fort Chadbourne were not reactivated after the war.

Between 1858-61, the Butterfield Overland Mail route crossed the ranch at a point now identified by two marble markers—one at a gap in the hills and another on the river. Across the river near the Camp Cooper site is the cabin which John Larn, the sheriff-turned-outlaw, built as a hideaway for him and his new wife, Mary Jane Matthews, one of Watt's aunts. Larn was arrested in 1878 by a posse that surrounded his home near the Clear Fork early one morning as he went out to milk his cow. It was known that for this chore he did not carry the pistol with which he was so proficient and so willingly used. Members of the posse included John Chadbourne Irwin and a number of other men from the region who had every reason to believe that Larn was stealing and slaughtering their cattle in order to fill his army contract quota for fresh beef destined for consumption by soldiers at Fort Griffin. Larn was later killed in jail in Albany by masked men—

men he thought had come to rescue him. The identity of his killers has remained shrouded in secrecy, but rumor has long sparked the story that Larn recognized at least some of the men among those who came to put an end to the outlaw's career. The widow later married John Brown.

Other of Watt's relatives figured significantly in the history of this area and in the larger West. Several members of the family had experience in trail driving, those often romanticized movements of herds of two or three thousand wild longhorn cattle to markets in Kansas or to breeding ranges on the Northern Plains. By the late 1870s, the seemingly boundless grasslands had been cleared of the vast herds of buffalo that once made the Plains seem a brown, moving mass. Earlier in 1865, the first herd of longhorn cattle was driven from the area by George Reynolds, Si Hough, and Riley St. John. This herd, originally headed for Old Mexico, was driven instead to Santa Fe in New Mexico Territory and turned a nice profit for the owners. In 1872, when he was only nineteen years of age, Watt's father was in charge of a herd of Clear Fork country cattle being driven to Nevada Territory on the Humboldt River. He had a long career as a cattleman and went on to serve two terms as County Judge of Shackelford County after he was first elected in 1894.

Among recollections of his mother's brother, Phin, are experiences on trail drives. There is little romance in his account. In one section he recalls a disaster involving over four thousand cattle being driven from near Goodnight, Texas, to the Dakota Ranch of the Reynolds Cattle Company. The cattle were split into two herds, one ahead of four different herds of the famous XIT Ranch and another following behind the XIT herds. Tot Richards of Throckmorton County and Tom Matthews were in charge of the two MO herds. Twenty-five miles north of Hugo, Colorado, misfortune struck on May 31,1892. A fierce, late-winter snowstorm so scattered the cattle and mixed them up that much time was lost as all the cowboys associated with the herds worked to separate the nearly fifteen thousand animals once again. The XIT Ranch lost three hundred head, the Reynolds herd only eight. However, the Reynold's crew lost twenty-eight head of horses in the bitter cold. The horses did not freeze to death, but the cowboys rode them hard and left them sweaty when the mounts were changed. Not in particularly good physical condition anyway, the animals simply chilled and died. Reynolds was forced to wire back to Albany for additional horses to be shipped by rail to the herd so that the cowboys would have a sufficient number of mounts to ride on the remainder of the journey.[6] These were indeed interesting and trying times in the frontier movement of the cattle, and little of the silver screen romance is seen here.

Another of Watt's uncles on his mother's side, Glenn Reynolds, moved to Arizona to ranch but was later elected sheriff of Gila County, Arizona. Phin recounts that Glenn was the only Democrat elected in that county up to that time. During his term, however, Glenn was brutally murdered in 1889 by a band of renegade Indians that he was transporting to the penitentiary at Yuma. The Apache Kid, an infamous fugitive during this period, was the leader of this group of renegades and killed the sheriff with the lawman's own weapon.[7]

More than cattle dominated the area of West Texas along the Clear Fork. As was true in many areas of the West, wild horses once roamed the area in great numbers. An unnamed mustanger recalls capturing the wild animals along California Creek, the area where Watt's mother and father were later to start their ranching operation. The method employed by the mustanger was to ride several of his grain-fed saddle horses in turn to keep the mustangs constantly on the move and thereby wear them out. He also wore himself out, and for what he considered very small return for his efforts. He discovered that the horses of true mustang lineage had very little value for they were small in size and mean in temperament. Escaped horses and mules, especially government stock, often ran with the wild animals and were valuable but were often more trouble to catch than they were worth in markets such as those in Fort Worth.[8] This mustanger's career was brief.

Expanding influences of civilization were felt in the area after the Civil War when protection returned in the form of Negro soldiers in Companies F, I, K, and L of the Sixth Cavalry. These were the "buffalo soldiers," so-called by the Indians because of the troopers' curly hair. Later, units of the Seventeenth Infantry joined the contingent. The soldiers were garrisoned at the newly established Fort Griffin, which was built on the bluff overlooking the Clear Fork bottom land after 1867 and prompted the development of a town by the same name which grew up on the flat along the river.

The town of Fort Griffin was as rowdy and dangerous as any other on the frontier. In addition to the soldiers, a sizeable contingent of buffalo hunters outfitted there for the summers of 1877 and 1878, the two years that saw thousands of "flint" or hardened green hides come back through the town of Fort Griffin and also saw the demise of the buffalo herds on the Texas Plains under the guns and skinning knives of a veritable army of hide hunters. In 1877 alone, over two hundred thousand hides came through Fort Griffin. Merchants there sold tons of lead and kegs of powder and other necessary supplies to the hunters before these men moved out onto the ranges to slaughter the buffalo. Joe McComb, John Poe, John Jacobs, Alex Howsley, and J. Wright Moar (who killed the only white buffalo known to have been killed in Texas)[9] worked out of Fort Griffin, and descendants of some of these early hunters remain today in the Albany area.

The town was also a hangout for various types of ne'er-do-wells. The infamous figures who frequented the town included Wyatt Earp, Pat Garrett, and Doc Holliday, along with John Larn and John Selman and less well-known figures such as Comanche Jim, Bill Henderson, and Charlie McBride. Selman left when John Larn was captured by a posse but lived long enough to kill John Wesley Hardin in El Paso years later. Prostitutes plied their age-old trade in the town as well. Big-Nose Kate, Hurricane Minnie, Sallie Watson, and a host of others left their legends there, as did the mysterious Lottie Deno, a lady gambler whose origin and final destiny are shrouded in mystery today.[10] Hangings were common, and murders even more frequent, as law-abiding citizens and outlaws battled to see which side would gain control. Thievery of livestock was rampant, and horse thieves were sometimes found hanged by members of the Vigilance Committee, a group of citizens determined to drive the criminal element from the Fort Griffin area.[11]

Other historic sites in the area deserve to be mentioned. Along the south fence which serves as the border between Lambshead and the Nail Ranch ran the MacKenzie Trail, so called because its frequent traveler, Col. Ranald MacKenzie, used the route for his numerous forays against the Indians on the High Plains. After repeated efforts, the soldiers caught the Comanche and Kiowa Indians in winter camp in Palo Duro Canyon in 1874. Although some Indians were killed, their most significant loss was that their winter supplies, teepees, and horse herd were destroyed. Thus was removed the last advantage that gave the Indians the means to travel and hunt between raids on border settlements. MacKenzie's action effectively ended the depredations of these skillful and dauntless warriors. Among the Indians was Quanah Parker, the Quahadi Comanche chief who, realizing that the end of an era had come, finally led his people to a reservation in Oklahoma.

Quanah Parker was the half-white offspring of a white captive, Cynthia Ann Parker, who was seized as a child in 1836 and reared by the Indians. His father was Peta Nacona, himself a Comanche chief. In 1860 Cynthia Ann was captured by Texas Rangers led by Capt. L. S. Ross and soldiers from Company H, Second U.S. Cavalry, from Camp Cooper and returned to her family. According to John Chadbourne Irwin, on her way back to civilization, she was brought through the Clear Fork area, where he saw her.[12] Once back at a "home" that was neither familiar nor friendly, she pined away and died in what was to her by this time a "foreign" culture.

The MacKenzie Crossing on the Clear Fork is up the river from Lambshead on property presently belonging to the Nail Ranch, land once owned by the Monroe Cattle Company. George Reynolds, one of Watt's uncles, was at one time president of this company. Northeast of Fort Griffin, on the banks of the Clear Fork, a picket-style house was raised by W. H. Ledbetter, who was later to become the first judge of Throckmorton County. The family moved there after the shocking kidnapping in the early 1860s of their young son, John Calvin, by Comanches when the family lived at the Ledbetter Salt Works located south of present-day Albany. It was for a swim in the waters of the Clear Fork that the older brothers years later took a young man who came into town with an Indian trader. The Ledbetter brothers wanted to see if the brand accidentally put on the young John Calvin by an older brother was still visible. It was, and the Ledbetters thought that their long-lost son had returned. However, the man refused to accept his birthright and even rejected the Ledbetter name.[13] The picket house rests today in a park-like setting in downtown Albany, a reminder of this incident from the frontier era.

Early newspaper accounts of the region survive. In addition, Edgar Rye, an early-day newspaper man in the town of Fort Griffin, later in Albany, left his flamboyant record of these days in an interesting book called *The Quirt and the Spur.*[14] The richly informative and entertaining files of the *Fort Griffin Echo* were later transferred to Albany and serve as a record of that turbulent frontier town on the Clear Fork.

The end came to the town of Fort Griffin gradually but inevitably. When the soldiers were no longer needed to subdue Indians and left the fort on May 31, 1881, an important part of the economic base of the community disappeared. Shortly thereafter the rails of the Texas Central Railroad were laid through Albany rather than through the Fort Griffin area. Trail drivers had begun buying supplies in Albany anyway, since Frank Conrad had moved his supply house there, so that economic support had already eroded. The hide hunters were gone, chasing new herds that soon vanished as the bison almost became extinct in North America. The ne'er-do-wells that preyed on the fun-seekers left for more promising areas, and the pace of life in the town slowed to a halt. With its demise, a calmer mode of life grew up along the Clear Fork.

In time, better and more permanent residences were raised. With the railroad nearby, sawed lumber was available to be hauled in wagons from Albany, not from Fort Worth. Houses were also constructed of native limestone. Ranching grew, and the ranges came to be enclosed with barbed wire to control the livestock. Early fencing practices included running two or perhaps three wires around the exterior of the ranch. Later the practice of cross-fencing to create pastures of manageable size developed. Earthen tanks were dug out using teams of mules and fresnoes to catch the often skimpy rainfall and thus provide access to grazing land far from streams. By the end of the century, a noticeable level of polite culture and economic stability had come to the Fort Griffin area.

Watt's family tradition is rich and strong in this area and beyond. His brothers and sisters who lived to adulthood have had unusually long and productive lives, most of them living well into their nineties. Watt had six sisters and two brothers: **Annie Caroline**, born November 21, 1877, died as a child, of dysentery on May 24, 1881; **Mary Louise**, called May, was born May 14, 1880, and died on September 25, 1962, the wife of sixty-three years of Thomas Lindsay Blanton. Mr. Blanton was a judge in

Abilene and later served for almost twenty years in the Federal Congress. He returned to Albany and practiced law there until his death. The couple had five children: Thomas Lindsay, Jr., John Matthews, Anne Louise, Joseph Edwin, and William Watt. A brother, **Joseph Beck**, called Joe B., died in Boulder, Colorado, in 1977, after many years of assisting Watt in supervising the work on Lambshead. He had married Louise Webb on May 18, 1918; they had one child, John Alexander II. **Susan Elizabeth**, called Susette, the wife of Arthur P. Burns, lived from October 15, 1884, until September 6, 1971. During most of those years the family resided in Paul's Valley, Oklahoma. They had three children: Susanne Molloy, John Howard, and Betty. John, who has been heavily involved in the history of Lambshead Ranch, is responsible for the preparation of a book, *Summers at Lambshead*, an informal but informed account of family life on the ranch. His book carries as an appendix the genealogical chart of the children and grandchildren of his mother's brothers and sisters, certainly a worthwhile piece of information for those interested in the family and the source of much of the information given here. **Ethel**, born August 28, 1887, became the wife of Liston A. Casey. She died in 1985 after living many years in San Antonio. They had three children: Sallie Joe, Mary Caroline, and Watt Matthews, called Palo. A licensed veterinarian, he owns and runs the Phin Reynolds Ranch north of Albany rather than pursue his veterinary practice. **Lucile**, born July 2, 1890, married Dr. Harold H. Brittingham, who practiced medicine in the Boston, Massachusetts, area before moving the family to Cleveland, Ohio, where their five children were born: Thomas Evans, Sally, Mary, John Matthews, and Robert Scott. Dr. Brittingham died at the age of 42 in 1937, after which Lucile moved to San Antonio for two years and then to Fort Worth, where she has spent the rest of her life. The namesake of her mother, **Sallie Reynolds**, born January 10, 1893, married Ardon Berkeley Judd in October, 1920, and died in 1985. Her husband worked for Republic Supply in Houston. Two daughters, Sally Ann and Susan Berkeley and one son, Ardon, Jr., came from this union. Their son, Ardon, Jr., is a vice president with the Dresser Corporation in Washington D.C. A second brother, **John Alexander Matthews, Jr.**, born in 1874, died on December 22, 1908, of a ruptured appendix. Because of the contributions they made, these individuals and their progeny are important to the life of the region and to the larger community far outside the geographic area of their birth. Their longevity has become legendary.

Part of the Anglo culture brought to the region by the Reynolds and Matthews families included two of the necessary manifestations of civilization—religion and education. Watt's family was instrumental in the first with the construction of the Matthews Memorial Presbyterian Church, constructed in 1898 at the corner of South Second Street and Jacobs Street, not far from the courthouse in downtown Albany. A stone plaque on the west wall of the entrance cites Joseph Beck Matthews, Watt's grandfather, as the one honored by the name of the church. This house of worship was built under the direction of the Rev. French McAfee, a native of Pennsylvania and a graduate of Harvard College and of the Princeton Seminary. When he moved to this fledgling frontier community, he convinced the people there to build both a church and a school. The church still stands, carefully maintained, still regularly used for worship services and, because of its unique grace and charm, for numerous weddings.

Modern conveniences have, of course, been added to the structure. The most remarkable addition is a large pipe organ. The main pipes were imported from Holland, and others installed in the 1980s came from Germany. The only electrical parts of the organ are the lights and blower motor. The exposed pipes and keyboard sit prominently in the east end of the sanctuary and produce a dramatic visual as well as auditory effect on the congregation.

Mrs. Matthews wanted Watt to be the first child christened in the church, but her wish was thwarted when another child was christened first. However, Watt became the second child to experience this rite there. Later he become a deacon of the church and served in that capacity for many years.[15] Although Watt was not the first child christened in the church, his oldest sister, May, married Thomas Lindsay Blanton in the first wedding performed in this remarkable facility so important to the Matthews family. In 1940, almost two years after Mrs. Matthews' death, a hand-carved baptismal font was installed and dedicated to her memory. The building is of sound construction and has never cracked or suffered structural damage. At least part of the credit must go to J. M. Ackerman, who was in charge of the construction of both the church and the nearby Reynolds Presbyterian Academy.

Rev. McAfee also stimulated the members of the congregation to found this academy to educate the children in the community. Watt's family figured in the earliest efforts to accomplish this task. Among the seven original trustees were G. T. Reynolds, W. D. Reynolds, J. A. Matthews, and N. L. Bartholemew, all relatives of Watt in one way or another. The three men who signed the charter for the school were all brothers-in-law: G. T. Reynolds was the oldest brother of Watt's mother, J. A. Matthews was Watt's father, and N. L. Bartholemew married Watt's mother's sister, Susan Emily. These names appear regularly in historical accounts of the school, which record the grand opening of the Romanesque-style structure on January 10, 1899, after classes had begun the fall term in an old store building on Albany's Main Street the previous fall. Although it was affiliated with the Presbyterians, the initial program had a nondenominational atmosphere, because the local Baptist minister gave the benediction at the opening ceremonies. The course of study was rigorous, with higher mathematics, science, history, classical literature, and rhetoric required in the curriculum. The school ran from 1898 to 1905; the academy was revived, however, in 1908, and the next year became Reynolds College, an undergraduate institution that seemed to fit in nicely with Simmons College – later Hardin-Simmons University in Abilene. These two were joined by Stamford College, a Methodist institution of higher learning then located in nearby Stamford but later moved to Abilene and renamed McMurry College. Although the academy and later the college brought refinement with music and drama and other cultural offerings along with education to the area around Albany, it closed in either 1913 or 1914 because local residents sent their children off to school elsewhere and funding fell far short of operational needs. In addition, Rev. McAfee felt called to leave Albany to take another pastorate in Portland, Maine. When the college closed, Albany lost its chance to become a college town, and the structure was later removed.

This educational effort has not passed into obscurity. The history of the school has been recorded in a book entitled *Reynolds Presbyterian Academy*, authored by a former Albany teacher named Ruby McGill Dodge.[16] The book, published in 1960, is graced by illustrations by photographer Donald Blakeley. The foreword is by historian Rupert N. Richardson, one of the most renowned of

Texas and southwestern historians and a former president and long-time professor of history at Hardin-Simmons University. Students who attended the academy included Berta Hart Nance, who gained some reputation as a poet and whose poem "Cattle" serves as the introductory piece to the annual production of the *Fort Griffin Fandangle*, an outdoor musical drama in Albany. Familiar names also included Gentry, Conrad, Huskey, Campbell, Cauble, McComb, Herron, Newcomb, Bennett, Dodge, Irwin, Leech, and others along with many occurrences of names of members of the Matthews and Reynolds families.

This is the context and culture into which Watt Matthews was born in 1899 and which he has helped and been aided by in these more than ninety years. He took to this life like a colt takes to the limestone hills and fertile bottoms of the Clear Fork country, and he speaks glowingly of it in a way appropriate to the cattle and oil culture that it is.

CHAPTER II

THE YOUNG WATT, HIS EDUCATION AND VIEWS ON EDUCATION

To those who do not know Watt well, the low-key image and rustic simplicity of his lifestyle do not shout of the influence of education. Yet just a brief conversation leaves a deep impression that Watt is not only well-educated but quite well-informed on the issues of the day. His opportunities for training were not easily accessible, and the fact that he is as well-schooled as he is resulted from the dedication of his mother and his own personal drive. Like that of so many people of his generation, his education was erratic and sporadic. In only one instance before he entered college did he attend any school for two consecutive years, and he suffered interruptions of one and two years in his formal schooling. He is quick to say that he certainly does not recommend this pattern but that "it just happened to me, juggling around one way or the other, the way I had to do back then."

He began attending school before he was actually old enough. In 1904, when he was only five years of age, he convinced his mother to let him attend classes at the Reynolds Presbyterian Academy in Albany. She consented, since the academy was to be closed the next year. There he became acquainted with a number of other children from the area. One of his classmates was John

Bennett, with whom he shared one of the two-seated desks used at the school. Here was formed a deep friendship that lasted until Bennett's death in the mid-1980s. In his later years, Bennett worked for Watt as the caretaker for one of the fishing camps on the ranch, one which still bears Bennett's name. Following the year at the academy, Watt did not go to school because he was still only six years of age and was ineligible to enter the public schools. When he reached age seven, he attended first grade in the public schools of Albany in 1906–07. One of his first grade classmates was Frank Whitney, son of a local grocer. Since two of his sisters were by this time enrolled in the University of Texas, the family took a house in Austin and moved the children there for schooling. The two oldest children were in the university, but Sallie and J. A. were in high school; Lucile and Watt attended the Whitis School, a private one for the earlier grades.

Then tragedy struck the family: John Alexander Matthews, Jr., the namesake of his father, suffered a ruptured appendix and died on December 22, 1908, at the age of fourteen. Heartbroken, the family moved the children to Fort Worth, where Sallie and Watt attended St. Andrew's School the next year. Then they came back to the ranch, and for two years the children did not attend school at all.

The first year back at Lambshead saw little attention given to schooling for any of the children still living at home. During this time, Watt and his remaining brother, Joe B., worked alongside the cowboys on the ranch taking care of the cattle and sheep. This period provided crucial training for his life's work: at an impressionable age he was learning the ranching business from the ground up from the back of a horse.

Concern over the lack of progress in his formal education began to grow the next year, however, and some of his sisters tutored him in academic subjects so that he would get no further behind than he already was. Their strength was not in mathematics, and at this point, Watt worried that he had dropped far behind in computational skills. Consequently, he was moved back to Fort Worth and enrolled in Eleventh Ward School in 1912. Fortunately, the same woman who had taught him in elementary school was at Eleventh Ward, and his praise of her is still high: "There never was a finer primary teacher in the world than Mrs. Ben Wright." Through encouragement and special instruction, she helped him make up the background he had missed. The next year Watt attended Central High School for the spring term of 1913 and then for his next instruction enrolled in Sirus Private School, taught by what he terms a "Princeton man." Watt recalls, "He was hardboiled. He sent grades home every day, and if the grades were not up to a certain level, then he made us come to school on Saturday. He was tough!" Watt responded positively to that challenge and succeeded admirably.

Watt's father was deeply concerned with the education of his children. In fact, he had joined others in that area to build a school in Reynolds Bend in 1888, a structure which served several Matthews children. Mr. Matthews also wanted all of his children to attend college because he believed a college education would give them an edge in life. Although some of the children attended the University of Texas, Joe B. decided to enroll in Austin College, an outstanding private school in Sherman, Texas. His stay there, however, was brief, for he quickly lost interest in furthering his formal education. Watt recalls the abrupt method by which Joe B. ended an abbreviated college career. When his father came home one evening, he was surprised to see Joe B. sitting on the front porch with his feet propped up on the railing. The father, considerably dispirited by the turn of events, nevertheless honored his son's desire not to return to school. But the father knew that his son needed a life's work and took the legal step of having Joe's "disabilities" removed when the boy was about eighteen years of age. This action gave the young man full legal rights to conduct business but not the right to vote. Then the father put the young man in charge of the Zug Ranch near Albany. Joe B. took to this activity with remarkable alacrity and did so well that his father decided there was no need for the other children to attend college.

Watt's mother disagreed on this point, however. Although she had never been to college, she valued learning and wanted her youngest child to attend Princeton University. This desire had been created and cultivated by Rev. French McAfee, the Presbyterian minister who led in the founding of the Reynolds Presbyterian Academy and the Presbyterian church in Albany. Watt recalls that his father was not in favor of sending the youngster to college at all, but was willing to do so, especially with the kind of encouragement that the mother provided to both the father and

Watt. As Watt recalls, "She wanted me to go and kept talking about it until she got me wantin' to go."

Even after the decision to go to Princeton was made, Watt was a long way from being able to satisfy entrance requirements. In determined fashion, however, he set out to remedy that situation. From 1915–17, he attended San Antonio Academy, the only time he was able to go to the same school for more than one year. He prepared as best he could to make the transition to the formal, demanding atmosphere of an eastern university. Watt caught up in his study of mathematics and in his other subjects, but even as he was ready to move to Princeton, he discovered that he did not have the required amount of Latin. Undaunted, he moved to the Princeton area anyway and studied at a private tutoring school to prepare for the entrance examinations. Here he caught up on his Latin, and felt fortunate because only a few years previously the school had required, in addition to the knowledge of Latin, an extensive background in Greek. He learned no Greek.

He recalls that even with this extra preparation, he did not quite meet the entrance requirements and went before what was known as the Bicker Committee, a faculty group with the authority to make admission exceptions for those deemed worthy of entry. Watt recalls that they "put the quiz to me" and that his first answer caused them to laugh. He said that after that initial exchange, members of the group changed their attitudes toward him and allowed him to enter the School of Engineering. "They let me squeeze under the gate," he says with a twinkle in his eyes. Watt recalls that when he got to Princeton and walked across the campus, it was "a whopping change for me. The people there were strange to me, and I was strange to them." The young man from the Clear Fork was undaunted, however. He made his marks in the School of Engineering during this war year of 1918, working along with about thirty other students enrolled in the same course of study.

He returned home that first summer, even though the professors in the Engineering School insisted that he stay for summer classes. The faculty members wanted the engineering students to accelerate their training to graduate early in order to help in the war effort. Watt's response was definite: "I've got to go home and work on the ranch." He did and was more convinced than ever that he should live his life along the Clear Fork. He realized the value of continuing his education as well. When he returned to Princeton that fall, he transferred to the regular academic program to major in economics and politics, one of the combined fields of study available then.

He fondly recalls those years: "I went up there and although I was mostly cross grain with what I found, I stayed and finally got out in 1921. I wouldn't take anything for my four years up there." A clipping from the *New York Post* for June 21, 1921, is secured in one of the family scrapbooks. Here is a reminder that Watt was in the one hundred seventy-fourth commencement exercise at Princeton. He set the pattern for eight other offsprings of his brothers and sisters to attend this university. Evidence that his class appreciated him emerged in 1962 when they awarded him a Meritorious Achievement Award. Although he still praises the setting of Princeton for its natural beauty, he never lost his admiration for the Clear Fork country.

Education has been important to Watt because it was not always easy for him to get. His formal training was important, but his most respected and persistent teacher was his mother. Watt recalls that she made all of them master the Shorter Catechism, one of the basic guides to Presbyterianism. He also notes that "even though she never got past the fourth or fifth grade, she proved that you can help yourself by self study." She was to her son the epitome of the self-educated but productive person. Watt recalls, "She could beat me too bad to talk about on anything such as history or literature, anything except math and physics, and she never saw the inside of a high school, much less a college. She knew more about the stars than I do," he says. "Even though I took a course in astronomy at Princeton, she would get me up early in the morning, and we would look at the stars and talk about them and where they were at that time of year. She taught me much that I consider valuable. She also wrote well. The style of *Interwoven* shows that she had something on the ball. That book has become a classic."

Watt has maintained an interest in education by encouraging and rewarding youngsters who persist in a desire for learning. He once flew to Vanderbilt University for the graduation of a nephew, John Overton Brown. The young Brown belonged to the same fraternity as the son of Texas billionaire computer magnate, H. Ross Perot. At a reception following commencement,

Watt became acquainted with Perot, who later led in the controversial reform of Texas public education in 1985, and who ran unsuccessfully for President of the United States in 1992. Watt recalls being impressed at the time with Perot's perceptions. Watt also holds strong views on the value of education and kept informed on the upheaval over the Perot reforms. He said, "I agree with a whole lot of what has been done in improving the quality of public schools in this state. I am in favor of athletics and the band and all of that, but teaching children to read and write has become a sideshow, and that's wrong." He speaks favorably of the Ivy League schools as being the only amateur athletic conference left and still believes that they have their academic and athletic priorities in line. He labels as "ridiculous" the process of bringing athletes through college because of their physical skills and not requiring strong academic performance from them as well.

The education of Watt Matthews deserves part of the credit for the success he has had in ranching and in his other activities and in his generous, open attitude toward life. He has proved that the liberally educated man is capable of making a success in what he undertakes if he has the character and desire to succeed.

CHAPTER III

THE EARLY RANCHING YEARS: EARLY TIMES TO THE PRESENT

Watt's life has centered upon ranching, a cultural heritage important to him, and understandably so. The Reynolds and Matthews families were ranching people who found the Clear Fork country much to their liking but also spread out from there. In the early days of ranching, the outfits that controlled the land were those that happened to be grazing a particular piece of ground at the time. The term "cow camp" is appropriate, because when grass got short in one area, the herd of cattle and the remuda of saddle horses could simply be driven to an area that promised better grazing. "Cow hunts," literally long rides by men hunting for unbranded cattle, were the practice of the day. At roundup time, the cattle were gathered and separated by brands identifying ownership. One of the gatherings in 1878 by the Reynolds' riders netted a herd of some four thousand cattle to be sold to the famous trail driver and rancher, Charles Goodnight. Six motherless buffalo calves were gathered with the herd and carried by Goodnight to the famous JA Ranch in historic Palo Duro Canyon in the Panhandle of Texas. Goodnight later recounted to the Matthews family that these buffalo calves served as the breeding stock for the buffalo herd that he kept in later years.

Roundups to work the cattle soon became organized. In the *Fort Griffin Echo* of August 23, 1879,[1] appears a notice for the fall roundups in the Fort Griffin area. The contents of the list indicate the wide spread of ranches in the area at this early date: District I, at the M. O. Lynn Ranch, began on September 14. District II, the J. H. Huggins Ranch, began on September 25. District III, the Reynolds and Peters Ranch, began on October 1; and District IV, the B. B. Yarbrough Panther Creek Ranch, on September 1. District V, the E. P. Davis Ranch, started on September 20. District VI, the King Ranch on Caddo, September 12. In addition, the Greg Ranch on the Clear Fork began the roundup at the same time. Notice is included that District VII, which apparently included the Millet Brothers Ranch at the mouth of Miller Creek, was not represented at the last meeting of ranchers involved in the roundups. District VIII, at the Matthews and Reynolds Ranch at Paint Creek, began on September 25, and the drives at the mouth of Sweetwater Creek commenced at the same time. District IX, at the Dave Rider Ranch, had its beginning date on September 22. These roundups were broadly announced to draw a large number of men, because on the open range the cattle roamed long distances between gatherings. For spring roundups, the men often had to ride far to the south to find cattle that had strayed seeking shelter from fierce winter storms. The usual arrangement was for the men to gather cattle on a home range but also to send representatives to other ranches to identify brands and later drive strayed cattle back to the owners.

Various other problems faced ranchers in this day. Again in the *Fort Griffin Echo*,[2] on April 26, 1879, is a notice that owners of small herds of cattle objected to the methods employed by the larger cattlemen in the general roundups mentioned above. Since the small ranchers could not afford to be represented at all of these roundups, they felt, perhaps rightly so, that they were being cheated out of part of their cattle. This was a significant problem, one not easily solved at the time, but the methods changed as the culture matured.

Rustlers were a problem during those early years, and livestock associations formed to counter the threat. Originally organized at Graham, the Cattle Raisers Association of Northwest Texas, the forerunner of the Texas and Southwestern Cattle Raisers Association, met at Fort Griffin in 1878. Such associations enabled cattlemen to look after their interests and gave stability to a fledgling but rapidly growing industry.[3]

Later the severe droughts and devastating winters of 1885 and 1886 brought calamity and ruin to the ranch country across the West, and the Clear Fork area was no exception. This trying period saw an end of the use of longhorn cattle and open range grazing practices. Homestead laws made it necessary for ranchers to establish ownership rights to property, especially to reliable sources of water, and the need to improve the quality of cattle prompted ranchers to fence their property and begin to upgrade their breeding stock. A new day came to the industry, and ranchers continued to cope with new problems as well as with old.

Periods of drought remained a constant threat to the area, and Watt recalls that in 1917 and 1918 severe drought returned once again. In 1919, however, ample rainfall resulted in an enormous grain crop. It was at this time that Watt learned the lesson that all rural young men of the day learned – how to handle a bundle wagon during the hot summer harvest months. The bundles of grain were loaded on flatbed wagons and hauled to the thresher. Home from Princeton, he learned quickly that there was a right and a wrong way to stack the grain bundles on the wagon. He became aware that some types of bundled grain stalks, such as wheat, are slick and will easily slide off if improperly stacked. He says, "You don't have to reload a bundle wagon many times in summer heat to figure out how to keep those bundles from sliding off." During these days, Gene Pickard, a boyhood friend and later a successful Albany-area rancher, often worked alongside Watt, and as young men, the two bached on the ranch for two winters.

One of the major chores facing ranchers in those days was complying with the requirement to dip all cattle in an effort to control Texas fever ticks. These bloodsucking parasites were discovered to be a source of the devastating disease that struck cattle brought into the area. The imported animals, unlike the longhorn cattle, did not have natural immunity from the ailment. Kansas farmers had discovered to their dismay that something about the longhorn cattle coming north in trail herds caused their own domesticated cattle to sicken and die. Many theories were advanced, but it was some years after the problem became apparent that the source of the disease was discovered to be the fever

Paul Cameron Smith
'88

tick. The only effective way of coping with the threat was to dip all the cattle periodically in insecticide-laced water. This process had to be performed under the watchful eyes of state inspectors. To accomplish this job, ranchers constructed dipping vats from stone and cement. The vat had to be deep enough for a grown animal to be completely submerged and wide enough to accommodate the large horns on the mature cattle.

Since these vats were expensive to build and maintain, ranchers often cooperated by sharing the use of dipping vats. The crews drove the herds to pens at the vats and forced each animal off into the treated water one at a time. One man had to stand on the side of the vat and wield a pole to push each animal's head under the water in order to kill ticks on the heads and in the ears. He also might have to help a young calf swim the length of the vat. After climbing out of the vat, each animal was stopped for a time in a floored pen at the end of the vat to allow the water to drip off and run back into the vat. J. A. Matthews early established a reputation as a successful builder of dipping vats, and those on Lambshead were used by a number of area ranchers. Although the tick eradication program was time consuming and expensive for ranchers, the procedure succeeded in eliminating the fever ticks from the Texas ranges and allowed easier importation of the other breeds of cattle so important to the beef industry of the West.

As the range came under fence and the ranches were more permanently established in the area, a definite pattern for working cattle developed. The now-common practice of feeding cattle in the winter was unknown in those early days, and die-offs in the winter cold and summer droughts were not uncommon. Many a cowboy spent part of his time during the winter skinning out what was called "poverty hides," cattle that died due to weakened condition in the cold. It was better to retrieve at least that much from the animal than to let the coyotes have all of the carcass.

In the spring, however, cowboys gathered the cattle. These roundups were done usually by men riding horseback to gather the cattle in a net of hard-riding, shouting cowboys. Ethel Matthews Casey, Watt's sister, states in her small book, *Reminiscences*, that in early days, Lambshead Ranch was divided into only two pastures: the 15,000 acre Culver, which was named after one of the original owners, who was also a physician who cared for the family; and the Old Stone Ranch pasture of 25,000 acres, named after the structure Capt. Newton C. Givens had managed to erect on this stretch of grazing land in 1856.[4] Later, additional pastures were added by crossfencing: the William Reynolds Pasture, the Hay Ford Pasture, the Lambshead Pasture, the Overton Pasture, the Weighing Trap, the VH Pasture, and the Horse Pasture. As the pastures were cross-fenced in order to gain better control, roundups became better organized. Much of the work came to be done in pens rather than in corners of pastures where the herd was held in check while ropers dragged calves to the branding fire. Later the ropers were often replaced by branding tables and working chutes.

Even in these early days, the swapping of help from ranch to ranch was a common practice. As a young man, Jack Pate worked on the Phin Reynolds Ranch just north of Albany before having a full career of ranch work in the mountains of New Mexico around Las Vegas. He retired back to Albany in the late 1970s. Pate could remember that some of his early days were spent working at Lambshead. When asked to help out there, he would get up early, saddle one of the two horses he kept up, and put a lead rope on the other one. He rode the twenty or so miles to Lambshead before daylight, rode hard working cattle all day, and then saddled his fresh horse and rode home that night. He noted that it took a good horse to do that kind of traveling, but in those days he kept good horses and fed them well. Pate saw a sharp contrast between the climate in the cattle country near Albany and the cold, often treacherous winters in the mountains of New Mexico, where cattle and horses not accustomed to that weather could not survive.[5]

The life of the cowboy in West Texas has changed in many ways since Pate and Watt were young. Watt notes that his father always was more demanding of cowboys than Watt himself has been. Watt attributes his more moderate practice to a difference in management style and altered circumstance. As long as the work gets done, he sees no reason to concern himself with exactly how it is accomplished. One of the major changes Watt recalls is that in early days men had to leave the headquarters early enough to ride to the pasture they intended to work, often a distance of several miles. Now horses are hauled in trailers pulled by four-wheel drive pickups. After the cattle were gathered, if any needed to be moved to another location, these animals were driven rather than hauled in trucks, as they presently are. The long hours of a cowboy's prowling on horseback to look for cases of screwworms have disappeared since modern science has all but eliminated that threat. In contrast, in those earlier days every cowboy had to carry a container of worm medicine, usually in a boot top cut off and laced across the bottom. This handy carrier was tied to the saddle. The smells of Smear 62 and other remedies were noticeable on most cowboys in warm months. It was not uncommon for three or four cowboys to gather a small herd of cows and calves in a fenced corner and rope the calves that needed to be treated with this medicine. The elimination of this threat has done much to relieve the cowboy's routine work, for doctoring for screwworms was an ever-present job in the spring and summer when calves were small.

Other improvements in the technology associated with the cattle industry also made the cowboy's life easier. For instance, improvement in watering facilities has freed the cowboys from checking waterholes late on summer afternoons to look for cattle bogged in the mud of receding water supplies to which they had fled to escape the heat and the nagging bites of heel flies. Often a cowboy would be attacked by the animal rescued, although why cows pulled from near death would fight the men who had saved them is one of the mysteries of range life. The elimination of the fever tick, the use of sprays to control horn flies and other pests such as lice and grubs that cut into cattle productivity, and other changes have lightened the cowboy's burden and reduced the number of men necessary to do the work on modern ranches, all of which have contributed to a change in management style.

Despite these differences, the qualities of a good cowboy have not changed. Watt quotes his father, one of the best cowmen of the open range period, who said: "If you can't see a cow this year and know that cow and where you saw her last year, you are no cowman." When characterizing a good cowboy as opposed to a bad one, Watt says that a good one "can handle cattle without upsetting them; a bad one can't." According to him, there are still a lot of good cowboys these days; they simply do not get the same amount of practice that the old-timers did.

Watt has seen many fine cowboys come and go on Lambshead. One of the best known was Uncle George Newcomb, who rode for well over sixty years for the Matthews family. Born at Fort

Griffin in 1871, he became widely known for both his abilities and longevity. Two were black cowboys – Charlie Standifer, who served the family for many years as both a cowboy and later as a cook; and Tige Avery, in Watt's mind one of the finest of all cowboys he ever saw. Largely illiterate, Avery was an expert at reading brands and working cattle. He went on trail drives in early days, one with Phin Reynolds and a large herd of MO cattle in 1892 to Wyoming. As he grew older, he moved to town after several members of the Matthews family did in 1929. He still had noon meals with the family and worked as a janitor at the First National Bank for a time. He died in 1935 at the Texas Cowboy Reunion in Stamford. Since this was a busy time of year on the ranch, the family had the body buried at Stamford, with Watt reading the service. Watt is not prone to forget old friends, however, and later he had the body exhumed and brought back to Lambshead for its final resting place in the family cemetery at Reynolds Bend. The headstone reads simply "Bill 'Tige' Avery/ March 1853/July 1935/Fine Black Cowboy." The Matthews family honored Avery further by giving his name to an outstanding cutting horse which Watt rode in 1939 in Stamford at a cutting horse contest.

Trail driving Lambshead cattle to market was one of the oldtime cowboys' jobs, but the drive in Watt's lifetime has been far easier than during his father's career. Watt helped drive Lambshead cattle headed to market not to Kansas or other distant points but to the Albany railroad corral, a two-day trip from the ranch to the railroad siding. The ranch had a half section holding pasture on what is now the Palo Casey Ranch north of Albany, the one on which Jack Pate worked as a young man. There the men constructed a small building to serve as a camp house for the trail crews. The Lambshead cowboys would drive the cattle to that holding pasture and, accompanied by a chuckwagon, spend the night. There was no need to ride night herd, for the relatively small pasture was fenced. The next day the men would drive the cattle to a small holding trap adjacent to the railroad corral at Albany. There they had access to a set of pens and scales for weighing the animals. J. C. Irwin, Jr., remembers that about 1920 Judge Matthews paid him $5.00 to clear the brush off of that small tract of property, which today serves as the parking lot for the *Fort Griffin Fandangle* Prairie Theatre. Once the cattle were

loaded onto the train, the men would ride home, even if the ride lasted late into the night. Watt's perception of these changes is that the work of the cowboy has been reduced by well over one-half during the sixty or so years that he has seen the work done.

One of the most significant changes is in hauling cattle to market. The evolution of the practice on Lambshead is easily chronicled. The first trucking of cattle of market for Lambshead was done by Firl Irwin, himself the offspring of pioneer stock. His truck was a Model T bobtail rig. He could haul only a few head at a time, but the system was easier on man and beast than driving the animals to market. If only a few animals needed to be moved, it was more practical to use the truck. Gradually the use of trucks replaced the need to drive the cattle to the railroad corral, and thus began a practice that eventually evolved into trucking the cattle to the yards in Fort Worth and later to area auction rings or directly to feed lots in the grain-producing areas. Times change. Now the railroad tracks through Albany, once so instrumental in developing the cattle industry there, have been totally removed; and cattle are hauled to market in huge multi-level vans that ply the nation's highway network.

The breeding cattle on Lambshead have long been blooded Herefords, and for good reason. When the day of the longhorn was obviously over in the mid-1880s, the Matthews family and other ranchers experimented with various breeds but favored Shorthorns and Herefords. Albany bills itself as "Home of the Herefords," and the breed was introduced into Shackelford County in 1885 by Col. G. E. Waters.[6] Some of these early imported cattle were driven in from the coastal area but often fell victim to tick fever. Later, bulls were unloaded from ships and hauled in wagons or on trains to areas beyond the tick line to the north. Frequently, only bulls were brought in because of the difficulty and the expense in transportation. One of the first Hereford bulls brought to the area belonged to Henry Herron, a rancher and former peace officer in the area.

In 1911–1912, during one of the frequent droughts to hit the area, the Matthews family discovered that the greatest die-off had occurred among the Shorthorns and that Herefords seemed to be better hustlers on the drought stricken ranges. The commitment was made in 1912 to stay with the Herefords, and the family has not wavered from that decision since then. Bulls from the

Herron herd were joined by those from breeders named Hayden, Meacham, Sedwick, Bridwell, Vanier, and later Weldon Edwards and others. For example, in 1972 two hundred twenty-five fine Lambshead Hereford heifers joined two hundred heifers and twenty young bulls from South Dakota to make up a four hundred fifty head shipment of breeding stock exported to Hungary. These animals were air transported by Seaboard World Airlines in a stretched DC-8 aircraft from O'Hare Airport in an effort to upgrade beef stock in the European country, an effort which continued for several years.[7]

During his years of ranching, Watt has seen changes come to the Hereford business. He recalls how the breeders kept working toward shorter, stockier cattle. He recalls how this tendency led to the dwarfism that almost ended the Hereford breed as a desirable one for beef raisers. Watt has been determined to produce the kind of animals that feedlot operators want—tall, large-boned animals that add muscle when put on rich feed and produce the cuts of meat consumers prefer. Although he has helped promote show cattle, especially at stock shows in Fort Worth and Denver as well as those shown by youngsters in 4-H and FFA work, he has never been interested in promoting an unmarketable configuration of cattle. "To give a prize to an animal the industry would not buy is not fair," Watt says.

A most significant innovation to come to ranching in Watt's recollection was the advent of rural electrification. The lines were run to the nearby Overton Ranch before World War II, but the war effort ended expansion of the lines. It was 1946 before electricity came to Lambshead. For several years, the Matthews family had used acetylene gas for lights, a practice that the city of Albany had also followed for some time. Watt recalls that these lamps provided good illumination but that electricity did more to revolutionize life in the country than anything else because it brought with it refrigeration, water pumps, and numerous other labor-saving conveniences. Watt recalls that his father's idea of luxury was "sleeping between sheets and drinking ice water." Even though "luxury" is a relative term, the coming of electricity certainly provided a measure of comfort to life on remote ranches. Refrigeration then provided the ice that had been available only occasionally from such sources as Mr. Raymond G. Little, the "ice man" from Woodson, who made a run through the Lambshead country every other day through the hot summer months in the 1930s.[8] Ice was not used for refrigeration in these days, as Mr. Little recalls, but found its way mostly into the glasses of iced tea to relieve the thirst of hot cowboys and family members.

Although cattle have always been the mainstay of the ranch, there was a time during the Grover Cleveland administration that Judge Matthews had twelve thousand head of sheep on the property. They were worth about a dollar each at that time, and wool was worth six cents a pound. There was, moreover, considerable difficulty in raising sheep, because, as the old adage says, "It is impossible to run coyotes and sheep in the same pasture." The predators just proved too pesky for the sheep to be considered safe, and the Matthews family decided to commit their main emphasis to cattle instead. There is to this day, however, a small herd of sheep kept around the ranch house both to keep up the tradition and to keep down the grass and weeds that would otherwise grow rank in wet years to provide cover for snakes and rats. No lawnmowers are needed at the headquarters, a good example of how Watt has kept nature at work on this ranch where the past and the present merge almost imperceptibly into a setting that could be easily romanticized and perceived as unreal. The reality, however, is very much present, as a visit early in the mornings on days that cattle are to be worked will attest.

CHAPTER IV

LAMBSHEAD RANCH TODAY

Throughout his life Watt has maintained a simple, almost Spartan personal routine. A lifelong bachelor, he has continued to live in the bunkhouse at Lambshead Ranch, a practice he established long ago. He recalls that he made the move because he was often asked by his family to move temporarily to the bunkhouse to make room for visitors. So, early in his life, he made the move permanent and has stayed there ever since.

Watt's daily routine has long been predictable during the time that cattle workings are scheduled. Arising well before daylight, as he always does, he dresses in his faded denim jeans, workshirt, and well-worn cowboy boots handmade by James Leddy in Abilene, and goes to the cook shack. There the cook has the coffee ready and breakfast underway in order to serve the ranch hands and any visitors who might be present before daylight brightens the horizon.

Watt often uses these early mornings to write in the journal that he has kept for many years. Also at this time, he handles correspondence before outlining the day's work for his foreman, Terry Moberley. Before long the crew begins to gather. Wide-brimmed felt cowboy hats are hung on special racks, and spurs jingle as men pass through the serving line and fill plates with eggs, bacon, toast or biscuits, and gravy, and sip cups of steaming coffee. This part of ranch life is a long-standing tradition at Lambshead, an echo of the importance of breakfast to cowboy life. No longer, however, are the men seated around a campfire and a chuck wagon, exposed to the elements; instead they are enjoying the comfort of a unique structure which in cold weather is warmed by roaring fires in the large fireplaces at each end of the building. Thick stone walls keep the interior cool in the summer and insulate from the cold in winter.

As it has been since the inception of ranching in North America, the busiest times of year is the spring when the year's crop of calves is vaccinated and branded and the young bulls castrated. A typical spring working day occurred on May 27, 1987, when eight cowboys gathered for this task. Five of the cowboys came from the neighboring Nail Ranch: George Peacock, manager of the Nail, along with Sonny Edgar, Keith New, Bobby Williams, and Tunny Hollingsworth. These came as part of an old tradition – swapping help. They help on Lambshead; the Lambshead crew helps on the Nail. These men had risen early enough to do their chores, saddle their horses, load the animals into a large trailer, and travel in two pickups to Lambshead before daylight. Travel time is about thirty minutes, even though the two ranches share a common boundary. The men from Lambshead included foreman Terry Moberley, Richard Ballard, and Billy Gilbreath. Breakfast for the morning, thanks to the efforts of Bill Cauble, a fine cook as well as an excellent western artist, was whole wheat pancakes, scrambled eggs, sausage made from hogs slaughtered from the feral stock on the ranch, and all the trimmings. By 6:15 the cowboys had finished a hearty breakfast to help them through a long morning, and Watt and Terry had lined up the day's work. Aubrey Lange, a helicopter pilot who has been working for Watt part-time since 1981, was also there to lend his efforts to the work for the day.

The Nail men, their horses already saddled, were ready to ride. The Lambshead cowboys saddled their horses and loaded them into a second gooseneck trailer, and the two vehicles pulling trailers moved out to a spot some six miles from the headquarters where the drive would begin. The extra men rode in the back of one of the trucks. Terry Moberley took off with Aubrey

Lange in the helicopter to locate the cattle. After arriving at the designated spot, the men unloaded the horses, tightened their girths or "girts," and mounted, while they waited for Lange to drop Moberley at the spot so that the foreman could mount up and direct the drive on the ground. The conversation, storytelling, and news swapping thrived as it always does at times like this when the men have a few minutes of leisure.

The helicopter was soon buzzing across the landscape and before long landed to deposit Moberley, who mounted his horse and gave directions to the crew about how to spread out across the pasture and begin moving the cattle toward the pens. Soon Lange started driving the animals out of the remote corners of the pasture. During the drive, he skimmed low along the line of cattle to keep them moving toward the pens. Occasionally, Lange hovered over a particular spot and through his external loudspeaker called to a cowboy to come back and get behind a cow and young calf hidden in the brush. The bawling of the herd created a cacophony of noise as cows, separated from their calves, called constantly, seeking their offspring; and numerous young animals returned the calls. One small calf, only a few hours old, obviously could not make the drive, so two of the cowboys dismounted. One held the other's horse while the first caught the animal and carried it to one of the nearby gooseneck trailers to be hauled to the pens. The cow was driven on despite her repeated efforts to return to her calf. Had the trailer not been close, the men would have tied the calf's legs and returned later in a pickup to haul the young animal to the pens.

Cowboys dashed back and forth, shouting to urge the cattle on while dodging clumps of prickly pear. The riders often bent low or to one side to avoid low limbs and the thorns of the mesquite trees. The lush vegetation and the leaves on the mesquite trees offered ample protection for wily cows that wished to hide their calves from the drovers, so the men were forced to ride through the brush and also to rely on Lange's observation from above. The brightly-colored Herefords flickered through the brush, creating a montage of red and white against the shades of green of the landscape. After a drive lasting almost an hour, about one hundred twenty cows with their calves were gathered and penned. The horses were panting and dripping sweat; the men were perspiring profusely from the strenuous ride.

Surprisingly, this time no calves managed to escape the dragnet by rushing at the last minute under the bellies of the horses or through the fences. Typically one or two will manage to squeeze through the fence into the adjoining pasture. The cowboys must then find a gate to enter that pasture, rope the young renegades, and bring them back to the pens. Occasionally, a cow will decide to leave the drive. Once a cow succeeds in such a ploy, she becomes virtually impossible to pen. These outlaws must be roped and dragged into the pen. Often one cowboy will pull the cow by a rope around her neck while two others pass a rope around the back of her rear legs and maneuver her into the pen. This method avoids choking her excessively.

Once the animals were put into the trap adjoining the pens, Moberley rode among them to determine the condition of the herd. He found it to be excellent because of the rank vegetation in the pastures due to abundant rainfall, certainly unusual for this section of the country. On Moberley's signal, the herd was pushed into a smaller trap adjacent to the pens and then into the pens themselves. The cowboys dismounted, wrapped the bridle reins around the horses' necks in such a way that the animals could not step on the reins and thus break them, and moved in to separate the cattle on foot. Sometimes the horses are tied for this part of the operation, but two of these horses were known to be "rein breakers," animals that had developed the bad habit of pulling back on the bridle reins and breaking them when tied so that the animals can move about and graze. Once they are loose, the horses walk between the fence and the tied horses and catch bridle reins on saddle horns, thereby breaking the reins on other horses tied to the fence. Hobbles are rarely used by cowboys in this area.

Watt and Carroll Martin, his business manager, along with Donald White, joined the group at this time to assist in the working. The calves were separated from the cows in order that each might be worked as a group. First the cows were run through the chute and given an injection to kill grubs in their backs. The horns of all of the young cows were clipped so that the keen points would not injure the other animals or the men when penned in close quarters. Then the mature animals were moved into a pen where they were sprayed by three members of the crew — Peacock, New, and Williams. The sprayer used for this job was a large water tank, mounted on a two-wheel chassis, and the pressure was provided by a pump powered by a gasoline engine. The long hose was equipped with a nozzle which sprayed a heavy mist of insecticide-laced water over the cattle to kill the hornflies that bedevil the animals during the warm months, particularly in weather as damp as it had been that year, a condition conducive to infestations of flies.

The remainder of the crew began working the calves that needed to be branded, castrated, and vaccinated for black leg, a fatal illness for bovines. In addition, the male calves were given an injection of a growth stimulant at the base of the ear. The cattle were worked in this way: the working table, or "calf cradle," as it is sometimes called, was staked securely to the ground at the end of a long chute. The calves were put down the chute and individually run onto the table, which traps the animal and flips it onto its side. A moveable panel exposes the animal for the branding and, if necessary, the castrating. Watt himself wields the branding irons. These are heated in a propane-fired furnace fashioned from a piece of six-inch pipe with a plate welded over one end, a gap about ten inches long cut near that end, and a plate welded to the side of the pipe below the gap in order to hold the irons in the flame. This method is very efficient for keeping the irons hot. The AV brand is burned on with a V-shaped iron, which is first burned into the hides in the fashion of an A and then turned to form a V lying on its side. This pattern is the longtime and well-known brand of Lambshead Ranch. In addition, the heifer calves were given a year brand, a 7 since this was 1987. In this way, if the heifer is kept for breeding stock, the owners can cull the cows at the end of the ten-year period, the normal age at which cows reach the end of their highest level of productivity.

This part of the operation was completed at virtually the same time as was the spraying. At that point, a crew of three, led by Terry Moberley, went to another set of pens where Aubrey Lange had penned seventeen cows by himself, a feat which he accomplished with some ease by using his helicopter. The remaining cowboys loaded their horses into the trailers and drove to the bull pasture to gather the herd of bulls. In addition, ten part-longhorn yearlings had been gathered with the cattle in the morning drive, and these were loaded into one of the trailers in order to be hauled back to the headquarters to be shipped to market.

The gathering of the bulls was both interesting and dangerous.

Even though they were not around cows, where they are almost certain to fight, bulls are like teenage boys — they fight mainly for the fun of it and to prove their prowess. Each has his personal space and fights to defend it against any other bull, and sometimes mounted cowboys, that might intrude into it. Since the bulls were not with the cows, the males exhibited less tendency to fight, a happy circumstance for the cowboys, who must constantly be wary of bulls trampling whatever gets in the way of the battles for supremacy.

Lange, who had returned to the ranch headquarters, was notified by Watt upon his return to join in with the gathering of the bulls. By the time the crew reached the bull pasture, Lange was already airborne and was driving the animals from the remote parts of the pasture. The herd of bulls consisted of seventeen large and aggressive Angus bulls, nine young longhorn bulls, and twenty-six large Hereford bulls. The Angus stock is used for crossbreeding; the longhorns are kept for use with young heifers because the half-longhorn calves are slender in build and do not usually present a problem in being born in the pastures, away from any outside help. In this way, the yearlings can begin their productive cycle earlier than would otherwise be practical.

One of the Hereford bulls had a grass pack in his jaw and was noticeably dehydrated from the effects of his condition, for he could neither eat nor drink easily. He was very difficult to drive and required the attention, almost full time, of one or more drovers to keep him moving. The riders spread out across the pasture, which consisted of about fifteen hundred acres, and moved the animals toward the set of pens at the headquarters. Some were young and aggressive and ran quickly to the end of the pasture. The older bulls, quite surly because their period of peace and quiet had been interrupted, moved slowly. Finally, however, they were all moved through the gate. Unnoticed by the drovers, a gate to a nearby field had been left open, and Peacock and Ballard were forced to rush past part of the herd and drive eight bulls out of the field and back into the trap and toward the pens. Once the bulls were put into a large pen, the crew went to lunch. The time was 12:15 P.M. The morning had consisted of about six hours of work with very little break in the activity.

Lunch included thick slabs of sirloin steak cooked as a pot roast with onions, large chunks of deep-fried potatoes, green beans seasoned with bacon drippings, lettuce and tomatoes, and a marvelous cherry cobbler—all the work of Bill Cauble during the morning. Coffee was available, but iced tea was the preferred drink even though the temperature had not risen sharply during the unusually cool and cloudy morning the way it usually does on bright, sunshiny days in West Texas in late spring. Presiding over the meal and visiting with the hands was Watt himself. He was careful to inquire how Billy Gilbreath was feeling after a severe kick he had received from a cow while he was working in the pens that morning. In addition, Terry Moberley had also been kicked by a cow. Fortunately, however, none of the cowboys were seriously hurt or had been bucked off, and the morning was regarded as successful.

After lunch the men returned to work, penned the bulls, and began spraying them. Once they were sprayed, the bulls were run through the chute and given an injection to kill the parasitic grubs in their backs. Those younger bulls with sharp horns had the tips of those cut off, as had been the case with the young cows that morning. The men have to be very careful working around bulls because when two bulls are fighting, the one that is whipped will run away from his antagonist and trample whatever obstacle is before him as his enemy bears down on him to push the advantage to the limit.

The decision was made to try to remove the packed grass from the ailing bull's jaw. The animal was put into a chute and a bar passed behind him to keep him from backing up. Then he was roped around the horns, his head elevated, and another pipe inserted through the bars of the fence to hold his head up. The bull lunged against this pipe under this throat in such a way that he choked himself, perhaps from having dislodged part of the clot of grass. In any case, the bull died in the chute. Mortality among livestock is not uncommon in workings of this sort. Earlier in the week, five bull calves had died from becoming overheated and suffering the shock of being castrated. The deaths of these large calves cost the ranch more than $400 each, and the replacement cost of the bull might run over $1,000.

After the bulls were returned to the pasture, other minor chores remained to be done. The men separated the longhorn yearlings that had been brought in with the morning drive and hauled to the pens, as well as other stock that had been gathered over the past few days at the headquarters. These were sorted and weighed on the large scales at the corrals. Excitement erupted when one of the cows decided that she was being unnecessarily inconvenienced and began to try to run down any of the men she could reach. She was let out into a pasture adjacent to the pens, after a buffalo bull kept in that particular pen had been transferred to another in order to clear the route for the cow's release. The swift charges of the mature buffalo, desperate with fright and dangerous because of his size and speed, were a threat to which the cowboys cheered and shouted encouragement at each other as the bull was maneuvered into an out-of-the-way spot. In addition, three steers were cut out to be returned to pastures from which they had managed to escape, one to the Stone Ranch pasture and two smaller ones to the Reynolds pasture.

Once this work was completed, Ballard loaded the three steers into a trailer and hauled them to the selected locations while the other men returned with their horses to the area around the cook shack. The Nail cowboys loaded their horses into the trailer. Moberley unsaddled his horse and the one that Ballard had ridden that morning, and Gilbreath went out to wrangle the horses to get ready for the next day's working. He brought in about thirty head of horses and mules and proceeded to select the mounts he had been told to put up and feed overnight. Each man selected the horse by indicating the name of his choice to the wrangler. Moberley, using a hoolihan loop, caught his own horse and led him into a stall in order to feed him and keep him from being injured by the other horses overnight. Gilbreath opened doors on two stalls and called the desired horses by name. After some running back and forth, the horses selected entered the stalls and were fed. Once these selections were completed, the other horses in the remuda were fed and later released back to the horse pasture.

It is at times like this that oral tradition thrives among contemporary cowboys. Jokes are told, old stories remembered, farm and ranch hands discussed, and favorite horses recalled. Moberley told the story of a new pair of spurs made by Jerry D. Cates, a well-known maker in Amarillo. The price for the pair, which sported the name "Terry," was slightly less than $200; even though they were not so ornate as some the cowboys regularly wear when they work. This time of conversation is part of the social life of today's cowboy, and it is a time which the cowboys enjoy thoroughly.

The Nail crew returned to the home ranch, and two of the men loaded their horses into another trailer and went to check on a bull in the bull pasture. New and Hollingsworth returned to their camps to take care of the stock and chores that would last until well after 5:00 P.M. Arrangements had already been made for the same men to return the next day to repeat this routine with the variation found in the different pastures and pens.

The work on Lambshead is obviously modernized, but it retains the flavor and routine that has characterized ranch life from the beginning in this area. One important innovation is the use of the helicopter. Watt says that although he is not necessarily proud that he needs the helicopter in his roundup because the use is so modern, he recognizes that it does save him a lot of time and trouble. For example, a relatively small crew of men can work two pastures a day instead of the usual one, if Lange is there to help. Watt recalls that he first decided to use the helicopter in 1981 when his crew was having difficulty penning cattle in a pasture that was cut by numerous draws and some curious bends in the river. He arranged for the service of a helicopter from Big Spring, as he recalls, but at the last moment the pilot was unable to bring his craft. John Matthews had used Aubrey Lange at the Kickapoo Ranch, and Watt decided to call him.

Lange, who learned his helicopter skills flying two tours as a medical evacuation pilot during the furious action of the Vietnam War, recalls that the cattle he came to pen that first day had learned to take advantage of the deep draws and steep river banks to elude the cowboys. Therefore, he was unable to pen all of the cattle the first time, even with the cowboys' help. After the first drive, he returned by himself and herded the remaining cattle to a different set of pens on another side of the pasture and cleared the area in one drive. Lange observes that in large, rough pastures, he can gather more cattle than the rancher realizes are in that pasture. Thus, Lange pays for his services in extra cattle penned.

Watt says that in half a day the cowboys can do with Lange's help what it usually would take a week to do and is high in his praise of Lange, who he feels has the added skill of knowing cattle. Watt knows that no matter how skillful a pilot is, if he does not know how to handle cattle, he is useless. Lange is so good at what he does with his helicopter that he can pen the buffalo herd as well, a chore the men on horseback find a real challenge because the buffalo can turn far quicker than cattle and have less herding instinct. Besides being particularly adept at herding cattle and buffalo, Lange also aids in predator control. When Lange is in the region, he sometimes carries Watt or some of his hands aloft in the craft in an effort to help cut down on the coyote population. Flying low over the ranchland, the men can spot the coyotes, which are killed with a shotgun at close range. Even this method has not noticeably depleted the population of this particularly adaptable predator. Lange also assists in making game counts to help determine the number of deer to be harvested each season and has borne more than one photographer aloft for shots of wildlife.

Watt has long been dedicated to growing stocker cattle for beef production, but he has in recent years begun to do some crossbreeding. He has been given some Angus bulls by Matt Matthews and is using them to raise some Angus crosses or black baldies, as they are called. He is not committed to using crossbred stock as cows, and so is not backing off from the longtime commitment to purebred Herefords. Neither is he interested in some of the more recently introduced exotic breeds. Instead he feels that any time a rancher gets away from the British breeds, he is asking for trouble, because the packers just prefer stock from these reliable beef breeds.

Only a small horse herd is kept on Lambshead. Watt's father did not like horses and kept as few as possible. John Matthews said at the time the ranch decided to increase its horse herd, "That's a right about-face you've made here." Watt recalls that when he began raising horses, the ranch had no horses that young people on the ranch could ride, and he determined not to buy Shetland ponies. He managed to get a Welsh stallion imported from England and had Palo Casey buy some Mexican mares in order to raise some small horses. He said of these, "A kid needs a pony that a cowboy can ride and do something on and not be dragged off on fences and clotheslines. Shetlands are ornery as they can be. We needed something that they could get up on and go." He feels that he accomplished that with a cross between the Welsh stallion and the Mexican mares. Then Dr. Don Bell brought a fine Arabian "bred-in-the-purple" stallion, which Watt used for a time with the Welsh crosses. The horse developed an ailment which proved fatal. At that point, Watt came up with Cheyenne, a good-dispositioned Quarter Horse stallion. Watt is pleased with the crop of colts he gets each year, but he is well aware of how easy it is to become "horse poor." He recalls his father noted that one horse "would eat or destroy at least as much grass as three cows would. What they don't bite off, they will romp and stomp off," and Watt feels that advice is sound.

The concern for grass is universal with ranchers, because when the grass is gone, the rancher has to feed the stock. Many ranchers are forced to feed during the winter because of heavily grazing the pastures by overstocking, especially in dry years. Watt has avoided the problem by keeping his range sensibly stocked and assuring enough range forage to winter the cattle. He saves not only on feed bills but also on the labor and the expense of dispensing the feed. The crew feeds the cattle some hay in those times when ice and snow cover the ground and grazing is impossible or drought grips the land. Watt says that cattle can "get spoiled about that feed and not try to make a living." Watt keeps the calves over the winter and sells his annual crop as yearlings; these may be fed during the winter to spur growth. On the thousands of acres that make up the ranch, Watt usually runs about twelve to fifteen hundred head of cows. This runs the total to an annual peak of twenty-five hundred to three thousand head, counting the yearlings and calves.

Life on Lambshead today has its hectic times as well. The *Fandangle Sampler* held in the spring at Reynolds Bend is always a period of great activity because the natural amphitheater must be readied. In addition, numerous out-of-town guests stay at the ranch during this time, some of them coming in aircrafts that use one of the two landing strips on the ranch. One of the airstrips, the smaller one, is located just north of the headquarters; the other is near the river close to Buzzard Peak. The larger strip is a mile long. Watt has been told that the grass runway is suitable for twin engine, piston-driven planes but questionable for landing a jet. It is also the responsibility of the ranch crew to be sure that the mules and horses are ready to pull the stagecoach and various other pieces of rolling stock that are an integral part of the *Fandangle*. This process includes checking the wagons and harnesses and working the kinks out of the livestock, principally the mules, well before the show. Watt has been generous to allow visitors to tour the ranch and always manages to find someone, in years past often John Vandiver, to guide these tours, even in this busy time. He takes particular delight in conducting the tours himself, if he has time.

Obviously, there is more to running the ranch than branding cattle, and one area of demand is seeing to ample water supplies. Water conversation is crucial to a cattle operation. No well water is available for the upland pastures; and like other ranches in the region, Watt has had stock tanks hollowed out by bulldozers to catch runoff water. The tanks – not called ponds as the East Texas equivalent are – must be cleaned out periodically in dry periods to retain capacity and eliminate silt and mud that can bog cattle when the water level is low.

The value of the Clear Fork of the Brazos River to the ranch for both recreation use and stock water is immense. Unfortunately, the city of Abilene for several years dumped its treated sewage into the Clear Fork several miles above the ranch, a situation that polluted the stream. Watt has been instrumental in gaining court action to force Abilene officials to improve their sewage treatment facilities. The officials have insisted that the water is potable but have steadfastly refused to pump this treated water into Lake Fort Phantom Hill, the municipal water supply. Their refusal to reuse the water suggests their lack of faith in their system of treating the sewage. This is only one of the battles Watt has helped fight to keep the range along the Clear Fork in good condition.

The water supply for the headquarters comes from a large earthen tank and is gravity-fed to the buildings. The regional water board complained that one of the tanks nearby, constructed in 1912, is hazardous because water seeps through the gravel bed beneath the dam. Almost twenty-five inches of rain in the Albany flood of 1978 did not wash the dam out, and Watt has improved it considerably since then. He said that the board, however, is threatening him with a thousand dollar-a-day fine if he does not follow their advice on the dam. Actually, Watt would prefer a flood control dam to be built on the creek that runs beside the ranch headquarters because there is a danger of flooding from this stream bed.

Another conservation problem is the control of the mesquite brush and prickly pear that threaten these ranges. The story is that mesquites were brought to the area by mesquite beans eaten by trail herds from South Texas and deposited in this area in the rich manure, after the beans had passed through the cattle's digestive process. Every year the trees came further north. Another theory is that the trees have been in the area for long years, but the growth was controlled by thick sod and periodic range fires as well as by prairie dogs keeping their fields of view clear. Overgrazing allowed the plants ready opportunity to grow and spread. Regardless of the source of the infestation, the trees thrive in this area, and chemical and mechanical means have proved ineffective in eliminating the problem. One other way is to root out individual trees with a bulldozer equipped with a special blade. Any effort to control the mesquite is expensive but necessary in coping with the problem. Some success results from controlled burning, a practice whereby a selected portion of rangeland is purposely set on fire during a time when weather conditions are suitable—low winds and at least moderate moisture levels. Then the raging fire is usually controllable within the area bounded by fireguards previously cleared by bulldozers. These fires always have the potential of jumping the fireguards, especially if wind speeds increase beyond those predicted or if a coyote, rabbit, or quail catches fire and carries the flames across the cleared area. Then an uncontrolled blaze may sweep the surrounding countryside. Fire is always a danger, however, and is carefully watched by the cowboys and others, who have firefighting equipment at hand to combat the blaze if it breaks out of acceptable bounds. Only a shift in the wind kept the massive range fire of 1988, which ravaged half a million acres around Albany, from reaching Lambshead, though the inferno burned more than half of the neighboring Nail Ranch, which lost more than sixty miles of barbed-wire fence to the flames. It is no wonder that all smoke clouds are carefully checked and monitored by ranchers.

The fire does destroy the brush and scorch the prickly pear, yet it is highly likely that despite the procedure, these plants will sprout again. After the burn, cattle and wild game thrive on the fresh growth, which flourishes if sufficient rain falls on the scorched areas. In fact, cattle prefer to graze the recently burned areas because of the tender new growth and the absence of dead, coarse grasses. Legend tells that fires started by lightning and even by Indians burned the prairies off for centuries and kept the brush and cactus under control. Watt has been a leader in the use of controlled burns as well as bulldozing and grubbing to retard growth of moisture-robbing plants.

One possible source of relief from the mesquite is the extremely high demand for the wood for cooking purposes. Mesquite-broiled meats are a delicacy from New York to San Francisco, and charcoal processors are including mesquite chips in their product. The sale of this wood for cooking as well as for use in fireplaces has been a boon to ranchers, who can now sell the wood, or charge crews to cut up the trees into a saleable commodity. Revenues from the trees may ultimately rival other sources of income from the land.

One source that mesquite will not rival, however, is oil production, and the Clear Fork country is known for production from the Caddo and Mississippi formations. Although many oil wells and tank batteries dot the ranch, the income from this production goes to the families of Watt's brothers and sisters, rather than going back into the coffers for maintaining the cattle operation. Instead the ranching operation must rely on what revenue the cattle themselves can generate. "This situation has kept those families from calling on us for money, which has been fortunate for us in the long run," Watt says. The oil crunch in the mid-1980s hurt the economy of all of Texas, especially when the price of West Texas Intermediate Crude Oil fell to well under $15.00 a barrel, down from a high of near $40.00 early in the decade. As one rancher put it, "We used to think that we were cattle people. When the crash came, we found out that we were oil people. We had to learn how to make a living with cattle again." The good managers made the adjustment, but many others were unable to do so.

Thanks to Watt Matthews, Lambshead is a successful working ranch as well as a historical showplace. His interest in retaining the working nature of the ranch is not altruistic—he is a rancher making a living at his chosen trade. He runs a ranch that is well-known for the quality of the cattle it produces and the cowboys who work there. Above and beyond the economic reality, however, Watt is also cognizant of the historical significance of what he does, and he has made a noteworthy contribution in this area as well.

CHAPTER V

HISTORICAL AND ANIMAL PRESERVATION

A man of Watt's abilities and economic means would not be content simply to work cattle. "If I have a hobby," he says, "it has been to maintain and preserve the landmarks and traditions of the pioneers." Especially on Lambshead Ranch is this obvious. He goes on to say, "No one deserves a lot of credit for doing those things that he likes to do. That's why I don't think I deserve a lot of credit for the preservation work that I have done. I have long said that we can put up no greater monument to our family than to hand this ranch over intact to the succeeding generation, and I fully intend to do just that."

Credit has come to him for this work, and deservedly so, for he has restored structures and preserved wildlife in the area in a way that far exceeds what one would expect. The main restoration projects on the ranch have involved six structures: the Reynolds house; the Bartholemew house; the John House dugout; the school house built on the ranch in 1888; his most impressive accomplishment, the Stone Ranch; and his most recent, the Tecumseh Ranch House.

Construction began on the Barber Watkins Reynolds house, in what is called Reynolds Bend on the Clear Fork, on January 1,

1876. Watt calls it a "little Texas frontier house." It was in this house that John A. Matthews and Sallie Reynolds married on a snowy Christmas day almost a year later. The Reynolds house is a smaller version of the Matthews house, which was built on the Clear Fork several miles below Reynolds Bend. The Matthews house, on property owned by Ross Sloan in later years, burned in the later 1980s. The Reynolds house was inhabited when restoration started, but it needed a new roof, new floor, and new plaster on the walls. In addition, the floor joists had to be replaced. In an effort to preserve as much of the original structure as possible, Watt used the old joists as posts on the front porch of the house.

The lumber used in the original construction of the house was hauled by wagon over one hundred fifty miles from Eagle Ford near Dallas, the terminus of the railroad at that time. Since lumber was scarce and expensive, the walls of the house were constructed of native stone and are eighteen inches thick. The roof, of course, is of wood. Originally, a temporary wall separated the large room downstairs, and a similar one provided a divider for the large room upstairs. Since these were removed occasionally for parties and other gatherings, Watt simply did not have them put back in the house. The kitchen constitutes the north part of the first floor of the structure. Fireplaces stand at each end of the large living room downstairs. The original mantle still stands with the west fireplace, but the one on the east was fashioned to match the original one, which had decayed over the years.

A small brass plate attached to the north wall of the living room on the first floor shows the level the flood waters reached in the 1978 flood at Reynolds Bend. Part of the water damage in the house was to the oldest piece of furniture in the structure, an organ given to Watt's mother by a relative in Colorado. Watt recalls that the flood waters simply melted the glue in the joints in the organ and scattered the pieces across the floor. He remembers especially the ivory keys mixed in with the mud and silt. The organ was sent to a repairman in Canadian, Texas, who fortunately was able to restore the piece to its original condition.

For many years, in the living room hung two framed dolls given to Sallie Reynolds in 1869 by Tonkawa Indians in the area. A stroke of considerable good fortune occurred when the flood waters came within one-fourth inch of the doll's feet but did not damage the precious relics. Above the west fireplace is the original grant document, giving 160 acres of land to Watt's ancestors. Lettering reveals that this was in section 95, 3¼ miles from Camp Cooper. On a small cabinet also in the living room stand the shaving mugs of Watt's father and Mr. Sam Newcomb, an early-day settler whose diary recounts much of the earlier history of that area.

The grounds and other structures around the house still appear rustic but are modernized in many ways. The area sloping toward the river to the north serves as the setting for the annual *Sampler* of the *Fort Griffin Fandangle* performed in the spring. Its location in the pecan tree-studded bend of the river near a tall bluff is one of the most beautiful and serene in the region. There is no reason to wonder why the spot was selected. Stone fences run around part of the property. The fence is of two distinct types – a dry wall or stacked section which dates from about the time of the house; and two other sections held by mortar, which Watt had built. He says, "I thought that if those old-timers could build miles of that kind of fence, I could build a few feet of it." Watt recalls a bit of humor from a New England guest concerning the stone fences. As they drove through the area, she referred to them as "stone walls." Watt reminded her that they preferred the term "rock fences." She quipped, "Yes, I have heard of the famous general, 'Rock Fence' Jackson." Watt says with a twinkling smile that she got the best of him on that exchange.

Other structures stand near the Reynolds house – a guesthouse, a cabin, and some storage sheds. These sheds contain about two dozen pieces of rolling stock used in *Fandangle* productions each summer. Pens stand near the road leading to the house. At times, buffalo are kept there for visitors to see. In one of the open areas of the guest house is a long wooden table from the York Store in the old town of Fort Griffin. The table was a gift from Hallie Herron, who ran the store for many years. Not far from the Reynolds house is Hay Ford, the crossing on the Clear Fork used by suppliers of hay going to Camp Cooper in the days when that army post was in operation. A typical early-day ranch house from a later period also stands at that location.

The Bartholemew house, built by Nathan L. "Barney" Bartholemew, former Throckmorton County judge and an organizer of the First National Bank in Albany, was built in the summer of 1876 on the Clear Fork not a great distance from the Reynolds house. Lumber used for its construction was also transported by wagon from Eagle Ford. Lime for the mortar was prepared at the site, and the bones of a small dog belonging to Watt's mother were inadvertently included in the burned lime when the pet accidentally fell into the hot furnace. The house is the only one on the ranch with a basement, the result of the background of Mr. Bartholemew, a native of Connecticut. He just could not think of building a house without this underground facility.

Before its restoration, which took about sixteen months, the house was used as a granary, usually the last step for an old house before it collapses. However, a good sheet iron roof had been put on the house, thus saving the structure from the ravages of the elements. As much as possible was saved from the original structure. The stairway in the two-story house was restored using most of the original railing, but nearly all of the rounds had to be recreated on a lathe by Darrell Jeter of Albany. Some of the interior walls are original and are still unpainted; any new materials were painted. Modern plumbing was added later. This home stands today as ready for use as it was in that frontier era when it was constructed, even though Watt admits the job was arduous and the butt of many jokes because of the numerous problems that had to be solved in the process of restoration.

The dugout is a comparatively recent structure. Built by John House in the 1890s, the structure is set into a low hill overlooking the Clear Fork of the Brazos River bottom. The walls are built of native stone, rather than of sod or wood, which makes it somewhat unique among dugouts. This structure was restored to habitable condition in the 1970s. Watt recalls that the dugout, as originally constructed, was occupied as late as 1912 or 1913 and then fell into disrepair. Although historically important, the dugout is not nearly so old as the restored houses on the ranch. The stone fences, located adjacent to the structure, were built by a man named McFadden in the 1870s when he had a house nearby. The structure has been obliterated by time. A small clump of mesquite bushes marks the spot where the house stood.

The schoolhouse in Reynolds Bend was built in 1888 by heads of families living in the area. Those who were involved include Benjamin Franklin Reynolds, Mart Dixon, Mart Gentry, and John

Paul Cameron Smith
'88

A. (Bud) Matthews. Miss Esther Hutton of Wisconsin was the first teacher. She lived in the house of one of the families and rode to school in a buckboard with some of the children. In pretty weather, the group walked. This same structure also served area families as a church. Of the schoolhouse, Watt says that it is as near a facsimile as he could arrange, since all original materials had disappeared over the years. The method of construction is 1 x 12 inch pine planks raised vertically with the joints stripped with 1 x 4's, a popular method of construction during the period. It is on the same foundation as the original and is, as nearly as could be determined, the same size as the original. In planning the project, Watt consulted passages in *Interwoven* as well as several people including Joe B. Matthews, his brother, who could recall the structure from his student days there.

Public response to this project has been heartening. In 1985 *House Beautiful Magazine* ran pictures of the school with the children of area rancher Billy Green in the restored structure.[1] In addition, a picture of the interior of the reconstructed building appeared on the cover of the 1975 *Old-Timer*, a publication prepared by the Junior Historian Chapter at Albany High School. The quaintness of the school always attracts admiring remarks from visitors.

The Stone Ranch, however, is undoubtedly Watt's most spectacular accomplishment to date. Legend has it that in 1856, while stationed at Camp Cooper, Capt. Newton C. Givens founded the ranch and constructed the complex of houses and the stone corral along what became known as Walnut Creek in southwestern Throckmorton County, about twelve miles west of present Highway 283 and five miles west of the location of abandoned Camp Cooper. The site is located south of the Clear Fork of the Brazos River, as are ruins of the small bake oven ordered built by Col. Robert E. Lee or by his successor for the short-lived relocation of Camp Cooper. Buzzard Creek is near by on the south side of the river. The name of the ranch derives from the stone structures.

The story of the founding is that Givens may have intended the place as a lodge for hunters, since he was an ardent hunter himself, but he may also have intended to raise cattle in order to sell beef to the army, a situation he supposedly found illegal only after the ranch was completed. This story has some interesting but little known developments, as a sketch of Givens' life reveals.

According to Heitman's *Historical Register and Dictionary of the U.S. Army*, Givens was born in Kentucky, entered the U.S. Military Academy at West Point on July 1, 1841, and graduated on July 1, 1845. His earliest assignments included a march from Fort Washita, Indian Territory, to Fort Wayne, Indian Territory, in 1845–1846 and service at Fort Smith, Arkansas, in 1846, as well as at Fort Gibson the same year. He served in the war with Mexico from 1846 to 1848 and participated in the Battle of Buena Vista. For his role in the battle, he was decorated for "Gallant and Meritorious Conduct" and then was promoted to bvt. first lieutenant on February 23, 1847.

He returned to frontier duty that included a march to Los Angeles, California, in 1848–1849, and was moved back to Fort Croghan, Texas, in 1850–1852. On July 25, 1850, he was promoted to a first lieutenant with the Second Dragoons. He was back at Fort Croghan from 1852–1853 and participated in a march to Fort Belknap, Texas, in 1853–1854, and to Corpus Christi, Texas, in 1854. He was then assigned with Company I, Second U.S. Dragoons at Fort Belknap, Texas, in 1854–1855. At this time, Givens' career took a decided turn for the worse.[2]

Watt had long known that Givens was twice court-martialed. Through Rufus Garrett, a Fort Worth attorney and himself a West Point graduate, Watt made a concerted effort to discover the nature of charges brought against Givens, but with no success. New information brought to light in the fall of 1988 by B. J. Fisher of Abilene may help explain the military trials of this mysterious man and his part in constructing the Stone Ranch buildings.[3]

The manufactured materials in the buildings – oak planks and split shingles – are thought to have been diverted from the supplies sent to help construct Camp Cooper but may not have been from that source. At least two reasons exist for this conclusion. First, according to Mr. Fisher, little in the way of materials was sent to the construction efforts of these frontier posts by the army.[4] The structures at Camp Cooper were mostly tents anyway. Whoever built the house probably had to haul in these supplies by wagon. Further support for this generalization about Givens' not using military supplies comes from the fact that at the time of this construction, Givens was not on active duty in the area near the ranch with the Second Dragoons because of the terms of one of his two courts-martial.

On May 4, 1855, he had been arrested and ordered to face a

court-martial in Austin, Texas. The charge involved "conduct prejudiced to good order and discipline" and "conduct unbecoming an officer and a gentleman."[5] The evidence against him consisted of a letter, of which Givens was thought to be the author, published in the *Texas Gazette*, a newspaper in Austin.[6] The text was highly critical of Givens' commanding officer, Maj. Enoch Steen, Commander at Fort Belknap, where Givens was then assigned. The conflict stemmed from an incident in which Givens led a contingent of twenty soldiers to discover the identity of a party of raiding Indians.

Fisher summarizes a report to the Adjutant General in Washington dated November 17, 1854, which recounts that Lieutenant Givens "was in pursuit of a group of Southern Comanche who had run off a number of cattle and oxen. He reported that he proceeded with twenty men to the Clear Fork of the Brazos. He found the river very high and was unable to cross it. A Mr. Burnett came down the river and told Givens that he had tracked the Indians to a place where they had crossed the river a few miles down. A Mr. Garner, who had assisted in trailing the lost animals, swam the river and served as a guide for Givens. The soldiers came upon two oxen killed and portions of their flesh taken off, apparently for provisions. The carcass of another oxen was found thrown in the river but none of the flesh had been removed from this one." Givens continued to follow the trail of the Indians. When he discovered that the guilty Indians were a surprisingly large party of southern Comanches, he returned to Fort Belknap to seek more soldiers to "chastise" the Indians.[7] Major Steen refused to authorize the expedition, and Givens became angry at Steen's lack of action. This situation is reflected in the letter, which was signed "By one who knows" and seems to reflect the information known to Givens and emotion characteristic of the young lieutenant.

Givens was in Austin in July of 1855 for his trial. According to Holden, Givens gained the title to the Stone Ranch property on July 20 of that year.[8] On July 23, 1855, only three days later, during the testimony at the trial proceedings, he made several derogatory comments about Bvt. Maj. Gen. Persifor F. Smith, commander of the Department of Texas. Although eventually found innocent on the first set of charges, Givens was subjected to a second court-martial ordered by General Smith.[9] While Givens awaited the result of his first trial, his unit was transferred from Fort Belknap to Fort Riley, Kansas.[10] Givens remained at Fort Belknap but was ordered for his second court-martial to Carlisle Barracks, Pennsylvania. He was there by Christmas of 1855 because he wrote on December 26 of that year a letter to Col. C. A. May asking to be transferred to warmer climate for his health and noting that since his "act or neglect . . . [occurred] while serving" in Texas, he expected to be tried there.[11] No change was ordered and his court-martial convened in Pennsylvania on January 15, 1856. Givens was found guilty, suspended from the army for nine months, and reprimanded. It is not apparent that he served the full suspension, for he resumed duty with his company on May 15, 1856.[12] It is possible but unlikely that during this time of waiting for the result of his first trial he may have spent some of his time building the Stone Ranch complex. He had as well a period of suspension after January 15, 1856, and may have worked on the project then. If so, he traveled in the dead of winter, probably largely on horseback, to the ranch and worked in the worst part of the Texas winter. One cannot be certain that he was allowed the leisure for this work because Heitman states that Givens was "in arrest, suspended, and sick" from 1855–1857.[13] Whatever he did during this time cost him heavily in his health. By January 7, 1857, Givens reported for duty in Austin claiming to be ill.[14] Concern for his health had become a dominant factor in his life.

By November 24, 1857, he was in Galveston, too ill from tuberculosis and intestinal problems to travel. A Dr. Herd, who treated him as best he could, wrote the following: "Captain N. C. Givens, Second Regiment of the Dragoons, called on me for medical aid on the 14th of November . . . . I found him labouring under a diarreal affection associated with a high state of the gastroy intestinal mucous membrane. His liver is involved functionally to a considerable extent. I have given him such medical treatment as in my opinion he required. In view of the fact that his lungs are tuberculous, I have advised him to remain here until he feels his health more improved before he makes the futher attempt to join his post."[15] Givens was on sick leave of absence after June 15, 1858, until his death in San Antonio on March 9, 1859. His effects were turned over to his mother.[16]

Holden indicates that the cause of death was an accidental

gunshot wound suffered while hunting with Capt. Kirby Smith, who had been stationed earlier at Camp Cooper, but does not footnote a specific source.[17] Heitman, a reliable military source, indicates that Givens died that day, but gives no reason.[18] It is Heitman's pattern to give cause of death, if unusual, so the gunshot wound story may not be true. No mention of cause appears in a letter from Maj. Gen. D. E. Twiggs, then commander of the Department of Texas, to Col. S. Cooper of the Adjutants General Office in Washington, D.C. One would assume that if too ill to serve, Givens would not have been hunting and probably died of complications of tuberculosis.

The idea that Givens built the ranch structures while stationed at Camp Cooper is erroneous. Givens was never assigned to duty at Camp Cooper. Don Biggers, an early source on stories from the area, seems to deserve credit for this false assumption, and later writers on this subject simply picked up that error.[19] Camp Cooper was home to the Second U.S. Cavalry, an elite group created by Jefferson Davis in 1855 and staffed by many future southern general officers. Givens' assignment was Fort Belknap, where the Second U.S. Dragoons were stationed, approximately fifty miles to the northwest between present-day Graham and Newcastle. These two units have often been confused because of the similarity in the names. Legend and fact meld together in this story, and Givens' true role in the construction of the Stone Ranch may never be known for certain. Givens will also be remembered as the commander of a small garrison of troops who set fire to the buildings at Fort Phantom Hill in 1854, when that post, despised because of the scarcity of wood and water and the harsh climatic conditions, was ordered closed by the army.[20]

Regardless of details of construction, the complex of structures was built. In the years following, several families temporarily occupied the site. In 1865 a ranching outfit known by the names of Knox and Gardner lived there briefly. Then in 1866 the Barber Watkins Reynolds family and ranch crew moved there. The house, then the outermost on the Texas frontier with none between it and the New Mexico settlements, is described in *Interwoven* as having two large rooms about sixteen by twenty feet with a fireplace in each end of the house. Double oak doors opened to the north and south. A spacious entry hall served as a spare bedroom when needed. One room had a floor of planks, the other of stone into which were cut cattle brands of the region. Watt regrets that his mother mentioned the plank floor, because he had trouble getting it laid to suit him. A small two-room house to the north served as residence for the young men and boys. About two hundred yards northeast stands the sheep shed; the smokehouse stands almost due west of the main structure. Records show that this structure, important for preserving and storing meats, once held a huge store of salted buffalo tongues, a delicacy during the period. Other families living there included the M. V. Hoovers from about 1872–1875; and the Millets, a ranching outfit known for rough antics, moved in next. About the time this crew left in the winter of 1879, fire swept through the buildings, and with time, the stones tumbled down. Many area residents believe the Millets set the fire themselves because they were distressed that their efforts to control the range in that area were unsuccessful. The Millets founded the Hashknife Ranch, which still operates extensive holdings some miles north of Lambshead. The Stone Ranch property was transferred from the Givens estate to John H. Hancock of Austin and later was sold to the Reynolds and Matthews families in 1880. Watt's father gained control of the property in 1885.

Watt had long hoped to restore the structure but was unable to find a stone mason able to handle the job. He finally located in Kerrville a man who could do the rock work; and on May 12, 1983, he began the task of cleaning up the site and sorting the stones. Extra stone was taken from the nearby creek. The tumbled-down walls of the nearby corral were reconstructed smaller than the original version. The first rock was not laid until June 1, 1983, but Thanksgiving dinner was served in the structure that fall. To guide the project, Matthews followed the scant description recorded by his mother in her book and old photographs of the structure made before the roof timbers collapsed. From this data, he decided the height of the walls and the pitch of the roof. The stone floor in the east room and in the "dog trot" is the original one. Even the large oak doors were reconstructed. The original keystone, with the date 1856 cut into it, was placed years ago in the chimney of the cook shack at the Lambshead Ranch headquarters but was returned to the house to complete the restoration. In addition, Matthews reactivated the well by cleaning it out and replacing the stone top with one from the Merrick Davis home ranch. He has also decorated the interior with artifacts from the period. On the west wall of the east room is a quilt in a glass-covered frame. This artistically done covering is the one made for Susette by her grandmother. On a table in this same room are two soft buffalo robes. The marker of a child's grave is visible on the hill to the northeast of the house.

The stone house and outbuildings stand today as proudly as they did in 1856 when Captain Givens somehow managed to complete this most unusual group of stone structures. Two bronze plaques hanging on the walls of the house carry wording from *Interwoven*. A stone marker with bronze plaques citing the history of the area as well as date of construction and names of the members of the Matthews-Reynolds family stands on the northeast side of the house. Formal dedication was held on this site at the centennial of the Reynolds Cattle Company on May 5, 1984.

One other project that Matthews has planned but has not yet accomplished is the restoration of the bake oven at the second site of Camp Cooper. That site is on Lambshead Ranch, but Matthews is hampered by lack of knowledge of the exact construction of such an oven. He had hoped to complete the project before he started on the Stone Ranch complex of structures, but, he says, "The river cut us off" because of unusually heavy rains, and no one knows exactly how to restore the oven. It is far smaller than the larger multiple ovens at Fort Griffin. Clifton Caldwell, member of a prominent local ranching family and historian of note, brought out a man with experience on such items. No successful plan was made, and no work was done, however. Watt assumes that the oven needs a stone roof, but he has found no one able to give the right kind of help in this project, one that continues to frustrate him.

His zeal for preservation and restoration was challenged by a major hail storm that hit the Bend in 1992. The storm broke out many of the windows in the school house in Reynolds Bend and damaged other buildings on the ranch. True to his motivation, however, Watt saw that these were repaired to pristine condition.

As part of the on-going maintenance on the many buildings on the house, in 1993, with the help of a man named Babu Patel, originally from India, Watt had all of the fire boxes in the several buildings on the ranch restored to first class condition. Since the dried mesquite wood available in large quantities on the ranch serves as both a source of heat and of comfortable pleasure, Watt felt it necessary to keep the fire boxes in good shape and was fortunate to find the help of this experienced mason.

Watt's drive to restore structures in the area centered in 1993 on the house on Tecumseh Ranch. Located across the Clear Fork from Lambshead, this ranch is now operated by John Matthews, Watt's nephew. The house, which dates from 1875, was constructed by John Selman, who went on to become known as a gunfighter and the killer of John Wesley Hardin in El Paso. The house, according to Mrs. Matthews' comment in *Interwoven*, was L-shaped with three rooms and was not fully completed when her family moved in. They had workmen add a front porch, along with "cupboards and wardrobe," and plaster and paint the structure inside and out. The outside stucco was painted gray and "penciled in white."[21] The house was destroyed by a storm in 1903.

Watt intended a restoration much like that of Stone Ranch, but the contractor felt that the foundation was inadequate for the new structure. After careful excavation of the old foundation and sorting of some of the stones from the original house by hired hands, the contractor used heavy equipment to remove the old foundation and prepare for a new one. The house is to be constructed of stone and as nearly as possible to duplicate the original.

One of the interesting problems in this project was the installation of the septic tank system. Holes were dug for the installation of the tanks which, when inserted, had to be filled with water to avoid the possibility of their being floated out of the ground in the case of a heavy rain. The trouble of getting that much water to the site at the time was a significant difficulty, but once it was accomplished Watt discovered that the tanks were located too close to the foundation and all the work had to be undone and redone. It is simply one of those situations that plague a project, which nonetheless by the summer of 1993 was in full swing.

Watt's efforts at historical preservation have not gone unnoticed, for he has received several prizes and awards. His Merit Award from the National Cowboy Hall of Fame and a Texas Preservation Award from Governor Preston Smith are among those honors garnered. Among his highest honors are two—The Golden Spur Award given to him in 1981 by the Ranching Heritage Center at Texas Tech University; and the Ruth Lester Award, the state's highest award in historical preservation, given to him in 1984, in Lubbock, by the Texas State Historical Commission. Watt was unable to attend the latter award ceremony in person because he was hosting a party honoring the centennial anniversary of the Reynolds Cattle Company at the Stone Ranch that same day. Also in 1984 he received the Josiah Wheat Award given by the Texas State Historical Foundation for his efforts in historical preservation.

Yet another award came on June 30, 1987, when the O. P. Schnabel Award by Keep Texas Beautiful was presented to Watt at the Anatole Hotel in Dallas. The award is made to "recognize Texas citizens who have given long-standing service toward improving their community's environment and quality of life."[22] Bob and Nancy Green, area ranchers and friends, nominated Watt for this award, one that justly recognizes the contribution of one dedicated to keeping Texas not only beautiful but historically authentic. The same year saw Watt and three other prominent Texas ranchers – Bryant Edwards of Henrietta, Foy Proctor of Midland, and J. C. "Togo" Moorhouse of Benjamin – recognized at the most prestigious of the state's ranch rodeos, the Texas Ranch Roundup at Wichita Falls on August 15, 1987. That impressive collection of cowmen's combined experience represented over three hundred fifty years of ranching.[23]

In 1989, Watt received the Texas Chamber of Commerce Cultural Achievement Award, and at the Spring Commencement was awarded an honorary Doctor of Humanities degree by Hardin-Simmons University in Abilene. That fall he rode one of the ranch horses named Tuffy, Jr., around the parade route in downtown Abilene as Grand Marshall of the parade opening the West Texas Fair and Rodeo. The next year he won the National Preservation Honor Award from the National Trust for Historic Preservation and was runner-up for the Resident Conservation Rancher by the Texas Soil and Water Conservation District in Region V. Also in 1990, he was inducted into the Great Westerners Hall of Fame at the Cowboy Hall of Fame in Oklahoma City. In 1992, the Texas County Agricultural Agents Association named him Man of the Year. Watt was unable to attend the ceremony held in El Paso, so the award was accepted by Lawrence Winkler, now retired as the county agent for Shackelford County. In 1993, he served as marshall for the downtown nighttime parade to open the Western Heritage Classic, now the largest single event held in Abilene. For this event he did consent to ride in a carriage rather than upon a horse. He was accompanied by Mary Featherstone through the downtown parade route. In this same year he was also the subject of a very fine article in *Progressive Farmer* magazine. Rather than slow down his efforts as he has grown older, Watt has continued them, and honors have continued to come to him for what he has done.

When asked about these awards, Watt replies modestly that he is just pursuing his hobby, "preserving the landmarks and traditions of the pioneers." Obviously, Watt has enjoyed preserving the landmarks on the ranch that he has managed since the 1940s; and contrary to what he says, he does deserve an enormous amount of credit for his efforts.

Watt's historical preservation efforts proved contagious. Joe Blanton, a cousin, has now restored the Hoover house, which stands not far from the Reynolds house in Reynolds Bend. Watt's brother, Joe B., was born in this house. Hoover, a brother-in-law to Watt's father, moved from the Clear Fork country to Kansas to become a farmer. This restored house adds to the value of this historic area.

Watt is also chafing to restore the John Larn honeymoon cabin located across the Clear Fork on the Putnam Ranch, at the old Camp Cooper site. He has not been able to gain permission to do so, and the stone walls, now nearly roofless, are crumbling. The four flower beds, each in the shape of a heart, a club, a diamond, and a spade, are barely visible except in the droughty summers when the vegetation shrinks to reveal the outline.

Watt's preservation efforts have not been restricted to structures. He has carefully preserved on the ranch animal life reminiscent of pioneer days. One of these reminders is a herd of about fifteen buffalo. When the herd grows too large, excess animals are slaughtered and the meat used to feed the crew and visitors. Some of the buffalo bulls are castrated and raised as steers, but adequate breeding stock is always retained. Although he uses the animals for meat, Watt admits that they do not have the conformation to provide the cuts available from beef breeds of cattle. The meat is not as high in quality or in fat as beef; but since the animals graze like cattle, the meat is very similar to beef. The buffalo hides are tanned by a taxidermist named Charlie Burdine and kept for future use. This herd has been on the ranch since about 1950, and some of the animals regularly make a trip to town to appear in the *Fort Griffin Fandangle.*

Maintaining this herd presents its own problems, however. Terry Moberley, ranch foreman, recounted that the favorite pastime of a buffalo bull is destroying wooden gates or lifting them off the hinges. Apparently, the animals like the challenge of hooking the wooden planks with their sturdy horns. Moberley admits that the bisons are hard to keep on the home range and to herd, but these animals are somehow content to remain on Lambshead. It is not unknown today for buffalo from one of the private herds in the country to strike out on a migratory trek across the plains, breaking through fences as they desire. This herd is difficult to handle by the cowboys and is often penned by Aubrey Lange using his helicopter.

Watt also maintains a herd of longhorn cattle raised from stock bought from the herd at nearby Fort Griffin in 1972. These rangy creatures, along with the buffalo, add an aura of frontier mystique to the setting. These are descendants of those creatures his father and uncles drove to markets and ranges to the north but which lost favor after the mid-1880s. They are impressive reminders of an earlier day for Texas cattlemen.

Feral hogs, turkeys, and deer abound on the ranch. One of the ways Watt's generosity is evident is at hog killing time. On a cold winter day the crew gathers and kills as many of these animals as the men have trapped and put on feed. In a normal year, about twenty-five to thirty will be killed this way. Several of the carcasses are cured, and the hams, bacon, and sausage are used to feed the crew on the ranch. Some of the carcasses are also given to those from other ranches who help with the chore and to members of the crew, as well as to friends and acquaintances of the Matthews family. Watt recalls that in 1932, an unusually good year, about ninety of the good range-fat hogs were butchered, so this is an old tradition. J. C. Irwin, Jr., an early resident of the area, recalls being given a freshly slaughtered hog carcass when he happened by the ranch headquarters late one cold evening in the 1930s.[24] Only then it was Judge Matthews, rather than Watt, who insisted on the gift to this young neighbor man.

Hunting seasons were closed in the area during the 1930s because uncontrolled hunting threatened to eliminate all of the wild game from the region, which had been a favorite hunting ground for Indians. Now due to protective laws, large numbers of game animals roam the hills and valleys. Watt deserves some of the credit for this return, for he has long been a game conservationist. Over six hundred wild turkeys have been trapped by the Texas Game and Fish Commission on Lambshead alone and transplanted to other areas as far away as Colorado and Wyoming to replace natural populations killed off by overhunting and poor range conditions. Such trapping began in the late 1940s, at which time most of the birds were relocated within the state. Later, stock was traded with other states for other kinds of game birds and animals.

Early efforts to trap the turkeys depended upon the pull type of drop net. Later the drop net powered by an explosive charge was employed. Bert Iverson, a game warden in Shackelford County, helped with these operations and after his retirement continued to help with game conservation. When he left, John Vandiver, a game biologist who had worked with Iverson on many of the operations, replaced the former game warden as the conservationist on the ranch. After coming to work full-time in November of 1978, Vandiver was instrumental in placing mule deer and oudad sheep on the ranch. Unfortunately, the mule deer did not adjust to the area and simply disappeared. Eleven oudad sheep were placed on the ranch, and Vandiver recalls that one pair of the sheep bore offspring. Unfortunately, the coyotes killed the young animal, a common problem for sheep of any kind in this country. It must be admitted that these two experiments were failures.

Vandiver proved extremely helpful to the ranch because he was familiar with the flora and fauna found there. In addition to his responsibilities in game management, which included keeping records of game animals killed by hunters during the season, he guided groups of curious city dwellers who wanted to see a real West Texas ranch in operation. He also helped guide hunters and fishermen who came to the ranch annually to take advantage of Watt's hospitality. Vandiver's records indicate that in the 1986 hunting season forty small bucks and thirty does were killed. For many years the ranch did not allow large-antlered deer to be killed, for as Watt said, "We are still trying to breed up to a better quality of deer." Vandiver just explained to the hunters what kind of deer could be killed and the wishes were followed. Of the hunting and fishing on the ranch, Vandiver said, "Watt has fed and allowed people to hunt and fish here for years and has never charged any of them a penny. He likes people, and this is his way of showing it."[25]

Many ranches in the area sell hunting rights, some for as much as $1500 for each large-antlered deer killed. Watt refused to take this step for many years, but some limited hunting is now allowed to lease holders. This is one concession to the board of directors that Watt has made in order to generate additional revenues.

One of Vandiver's additional responsibilities was to feed the game in the winter. During the winter of 1986–87, for example, Vandiver fed between twelve and fourteen tons of shelled corn to the deer, turkeys, quail, and feral hogs on the ranch. This feeding accounts in part for the return of wild game to this region along the Clear Fork, which has a reputation for large-antlered deer. The game is fed in an area close to the river that Watt refers to as the "park," located near Bartholemew Crossing and the Bartholemew house near Reynolds Bend.

The restored structures and the herds of animals constitute a mute testimony to Watt's interest in the history represented by his family's contribution to the West. He has been mindful of a more elusive kind of historical material as well. Although he has not contributed articles or books on the history of the Clear Fork country, Watt has seen much of that history as it developed in recent decades and has carefully preserved some of it in the journal or diary that he has kept for many years. In the first of these books he recorded events from 1920–1925. This set of records contains an entry of only one line per day in which he noted items of particular interest. In 1951, however, he began his present series in small calendar books. Standard form for each entry contained in these records includes the high and low temperatures at Lambshead and at Wichita Falls, the major meteorological reporting center to the north. He includes, as well, rainfall amounts, certainly important information to the cattle operation on the ranch, but he records items other than routine. The names of visitors to the ranch, who over the years have included important personages, are found here. Trips made to conventions, stock shows, and rodeos as well as births and deaths are noted. At one time I inquired of him some details of an honorary doctoral degree by Hardin-Simmons University to *Fandangle* creator Robert Nail. Watt was able to verify the date and the kind of degree given by referring to the entry in his journal. One of the most interesting details is found in the rainfall amounts of the 1978 flood—24.86 inches of rain at the headquarters and 33 inches at the river. Here is a man of the same orientation as the seventeenth century Englishman Samuel Pepys, who has prepared a record over the years which stands as an important account of life in this region.

When asked what he wishes he might have done in his life up to this point, Watt is philosophical. He says, "When I think about things that I might have done, I dwell on what I could and should have done for my parents. Too often we don't think about them until it is too late to do anything for them." Because of what Watt has accomplished for his parents and others, the present and future generations will have him to thank for the many accomplishments he has made in preserving a rich, historic past along the Clear Fork for the edification of us all.

CHAPTER VI

FORT GRIFFIN FANDANGLE

The town of Albany is known for many accomplishments, but without doubt the most famous of these is the *Fort Griffin Fandangle*. The play, a dramatic reproduction of the history, legend, and lore of the area, evokes a sense of the past that is difficult to describe. Each performance of this outdoor drama includes material on both civilian and military history of the area along the Clear Fork of the Brazos River near Lambshead Ranch. Watt has been associated with the *Fandangle* since its inception and has long supported it as a participant and a supplier of livestock and rolling stock. He has served for many years as chairman of the Livestock Committee and has long been president of the board of directors. Watt was in virtually every performance for many years and is due a special note of thanks for providing the setting of the spring membership party held in the naturally beautiful setting of Reynolds Bend.

Those thousands who attend the six regular performances each summer enjoy a special atmosphere as they sit in an outdoor amphitheater near Albany in late June. It is sundown, and the audience is talking excitedly. The throaty notes of a steam calliope boom the tune of a rousing song in the background. As darkness approaches, the spotlights click on, a lone soldier mounted on a horse rides to the low hill facing the audience and places a bugle to his lips, and the eerie sound of an organ starts a chill up the backs of the viewers. The words coming through the speakers highlight the tone of the evening by quoting "Cattle," a poem by Albany native Berta Hart Nance. The last couplet of this poem states the general impact of livestock on this region:

> Other states were made or born,
> Texas grew from hide and horn.[1]

And then follow the words "Ladies and Gentlemen, the *Fort Griffin Fandangle* of 19__." As the speaker ceases talking, riders bearing flags race across the grassy stage, and the spine-tingling figure-eight maneuver is executed at breakneck speed as the music of "Trail to Mexico" rises in tempo and volume. In dry years dust swirls, in wet ones mud flies, but the speed never slackens. These are not town boys riding borrowed horses; these are men who that very day chased cattle out of the brush along the Clear Fork of the Brazos River or branded calves either in the branding pen or by roping the heels and dragging the calves to the branding fire. This is the real West tamed just enough to bring it to town on toughened cowhorses that like the thrill of the headlong dash and seem to enjoy seeing how close they can come to the other horses that are expected to get out of the way. Only two horses have been killed in this spectacle in recent years, but one occasionally causes a spill. The riders expect that: this is the *Fort Griffin Fandangle*, and its purpose is to recreate the spirit of the frontier. It is perhaps well characterized in this rousing horseback ride.

The development of the material in the show is chronological, beginning before the habitation of the region by Indians. The opening scene depicts flora of the region – cacti, mesquite, bear grass, and others forms – by having a child carry onto the stage a painted prop depicting each plant or some stalks of tall grass. The same is true of the fauna – skunk, paisano or roadrunner, coon, deer, and yes, the mighty bison called the buffalo, Lord of the Plains – each represented by a child appropriately costumed. The intricate dance steps of these actors are an integral part of the show, and laughter ripples through the crowd as the child in a costume representing the rabbit "hops" across the grassy stage.

In order to take the audience back in time, the story is dramatized through the memory of the Old Timer, a character played by an old-time cowman recognized by area residents as one of the earliest still living. George Newcomb and then Joe B. Matthews played the role for years. One year Watt played the role. For several years Jack Pate rode a "herdwise pony" to the "ranch house by the creek" to tell the tales of the old time. In 1993, Tunny Hollingsworth, a cowboy retired after thirty-five years of working on the Nail Ranch, played the role.

The first settlement in the area was established by the Indians. Mounted actors, painted to fit the role, burst onto the stage riding bareback on spirited ponies, thus representing the Kiowa and Comanche bands which roamed the area, mounted on fleet horses that made them lords over their land. Dependent upon the buffalo, which furnished flesh, bone, and skins, the latter providing the tipi cover, these people lived close to the land and adapted to the area. When eaten raw, the meat of the buffalo provided a complete diet which was nonetheless supplemented with nuts, berries, and various plant roots when available. Though not sumptuous, the resources of the area were fully accommodating to the nomadic pattern of Indian life. The scene depicting this era features a tipi erected by the cast before the audience's eyes, and later a make-believe "buffalo" is carried in on a pole between two warriors. The Indians converge on the carcass, and when they withdraw a few minutes later, the stark bare bones are left lying on the ground. The heart is revered and carried off, an action stemming from the Indian belief that by this act of homage to the buffalo's key organ, the buffalo would remain plentiful. This scene celebrates the height of the Plains Indians' culture when they had the buffalo and the horse and no Anglo intervention.

One of the most remarkable openings ever used for the show was the appearance of a white horse ridden at breakneck speed by a cast member wearing an Indian costume. The mounted Indian appeared on a hill about three quarters of a mile in front of the audience. The spotlight followed horse and rider, mane and feathered headdress flowing in the breeze. The viewer could surely feel the chill of knowing the vulnerability of the settlers to the mounted, able Comanche warrior.

The calm continuity of this Indian life, interrupted only by tribal warfare, was shattered by Anglos migrating from the East, although the first travelers appeared harmless enough, for they were only trappers and early cowmen or travelers headed for the California gold fields. Mr. C. A. Bales portrays Jesse Stem, earliest settler in the area, as he leads his burro loaded with appropriate gear across the stage. These early settlers simply foreshadowed the masses that would come later to overwhelm the Indians and bring about the near elimination of the buffalo in a slaughter of unparalleled proportion. A song, "Let's Settle in This Country," emphasizes the determination of the settlers to establish homes here. The song "Dugout" written by James Ball, one of several talented citizens who have made a major contribution to the show, highlights the kind of habitation of this early day. A piece introduced in 1984, "The Stone Ranch," written by Louann George, highlights the historic structure erected by Captain Givens in 1856.

The Indians saw their way of life threatened and fought back in the bloody Indian wars. Although these conflicts are often romanticized in fiction and film, they were real, portrayed here in a reenactment of such raids as the one at Elm Creek and the kidnapping of Millie Durgan, and also the kidnapping of John Calvin Ledbetter at the Ledbetter Salt Works south of present-day Albany. This action sets the stage for the appearance of one of the South's most beloved men, Robert E. Lee, who commanded a contingent of soldiers at Camp Cooper from 1856 to 1857. He termed this post his "Texas home," despite his disparaging comments about the semiarid land and his longing for the lush vegetation of his Virginia home. During the Civil War, many of the Clear Fork settlers "forted up" in a series of buildings at nearby Fort Davis. When hostilities ceased and Union soldiers came back to guard the frontier, Fort Griffin was built to provide army protection for the settlers. In the production, the soldiers are white, not the black or "buffalo soldiers" actually stationed at Fort Griffin.

The military post which gave the name to the area was located on a hill overlooking the Clear Fork. Beneath it, on the flat in the river bottom, grew up the town of Fort Griffin, and it is this town that is the heart of the *Fandangle.* In song and dance, the life at the fort and town is dramatized. The gamblers, buffalo hunters, camp followers, ruffians, and other types are included. Actors represent historically accurate figures such as Lottie Deno, Johnny Golden, John Wesley Hardin, John Selman, John Larn, Doc Holliday, Wyatt Earp, and Big-Nose Kate; all of these plus others appear as the saloon scenes dominate the set.[2] Central to all the action in this part of the show is the Beehive Saloon, where these famous lines were lettered on the front:

Within this hive we are all alive,
Good whiskey makes us funny,
And if you are dry
Step in and try
The flavor of our honey.

In the midst of can-can dancers, fist fights, and gunfights, the raw life of the town of Fort Griffin is dramatized. "The Town of Fort Griffin" characterizes the place:

Oh, the town of Ft. Griffin,
So rough and fancy free,
The wildest town on the prairie—
A town that fits me to a T.[3]

The cattle drives occurred during this period of the 1870s, crossing the open plains east of present-day Dallas and, forced further westward each season, finally by Fort Griffin itself as the Western Cattle Trail traced its way from South Texas up the Leon River, on by Fort Griffin, and then across the Red River on its way to Dodge City and the railroad access to markets in the East.

One scene in this segment always inspiring awe, especially in newcomers, is the driving out onto the open stage of a herd of twenty to thirty outstanding specimens of full-grown longhorn steers from the state herd at nearby Fort Griffin State Park and other such herds. These rangy beasts, direct descendants of those which made up the herds that gave rise to the cattle kingdom, usually experience an exuberance for life sparked by the cooling temperatures down to about eighty-five degrees by this time of the evening. From the side, these creatures are awesome, standing six feet and more in height at the shoulder, their horns reaching out to six feet or more, occasionally even as much as nine feet. But, as J. Frank Dobie says, from the rear the creature shows "cat hams, narrow hips, and a ridgepole kind of backbone."[4] As the animals come down the incline onto the stage, inexperienced members of the audience gasp in fright, but the mounted cowboys hold the herd while one member of the cast, played for a long time now by Clifford Teinert, sings "The Cowboy's Prayer." When it is time to move the herd out again to a corral hidden in the mesquite brush over the ridge, the cowboys start riding in on the herd. Spotlights aim high to give a false light in order that the cowboys can see how to move the cattle. One rider turns and rides up the incline, and the herd leaves as if on signal. The animals snort and kick up their heels in the exit of one of the most exciting scenes in this or any other drama.

The cowboy life that grew up around these Spanish cattle is emphasized by a camp scene in which the cowhands play a trick on the trail cook, busy preparing his meal off the gate of his chuck box mounted on a covered wagon. A fire burns beneath an iron pot and the traditional enameled coffee pot. A Dutch oven sits ready. A song, "Come and Get It," rings in the night: "Come and get it, hear him shout,/Come before I throw it out." Then released onto the stage is a calf, hotly pursued by a cowboy – usually Bud Leech or Benny Peacock – swinging his lariat. The calf is roped, thrown, and "branded" right on the stage. The puff of yellowish smoke from the hot iron is obviously not a stage prop, but it comes from the hot iron applied to a piece of leather, not to the calf's side.

On a less serious side is one always well-received scene depicting a confrontation between a drunk cowboy returning to the ranch from a binge in town and a huge, thirsty rattlesnake with a yearning for alcohol and a knack for playing solitaire in the moonlight. Through use of a cleverly constructed stage prop, the cowboy and the snake share a drink or two while the snake watches the cowboy play cards. The cowboy decides that the association is an unnatural one and tries, unsuccessfully, to elude the snake, which constantly follows him from then on. This scene is a fine example of the humor present in the show synthesizing two of the best-known figures in the area, cowboys and rattlesnakes.

The heyday of the wild, open range period is thus depicted, but then civilization begins to close in. Into this turmoil of activity rattles the Butterfield Overland Mail Stage Coach, linking this outpost with the rest of the world on a route that runs from St. Louis to San Francisco. As settlement progresses, families are established, schools and churches spring up, permanent houses rise, and fences are built, with songs and dances for each phase of activity.

Then the death knell of the old ways sounds as the night is shattered by the screaming whistle of the machine that doomed Fort Griffin – the Texas Central Railroad train – which followed tracks laid into Albany, leaving Griffin to dwindle away as the trade left and moved to the other city. Thus, the unique ranch culture in the region is left to its own development.

An interesting incident concerning this railroad occurred when the football field in Albany served as the theatre. The real train ran on a schedule which always interrupted the show while passing on the tracks behind the stadium. To offset the distraction, the directors arranged for the train to become part of the show. Actors boarded the train, which timed its arrival at an appropriate point in the play. This is just another example of how the local situation was fitted to the needs of the *Fandangle*.

The finale of the piece is always the same – the entire cast standing on the stage and singing "Prairie Land," the favorite piece written by creator Bob Nail. It captures the essence of the frontier spirit of those who came to the region and stayed:

> Each man holds within his heart
> The dream of a land all his own.
> No other land will do,
> No other land is home
> Here beneath the Western sun –
> I know at last my quest is done.[5]

ALBANY'S RESPONSE TO THE OCCASION

No one doubts that the *Fandangle* put Albany on the map for many people; and for an event like this, the town shows its gratitude by staging a parade each year on the second Thursday of the performances. Storekeepers along the main street of Albany line windows with memorabilia from the frontier days – books, tools, items of clothing, and the like. The parade itself is usually led by the Hardin-Simmons University Six White Horses, each rider carrying one of the flags that has flown over Texas. Covered wagons rumble along the asphalt street, and an excellent replica of the Butterfield Stage rattles by, drawn by six feisty mules fighting to tear loose and run out their energy. Cowboys ride by, their hats pulled down and their horses prancing, frightened by the crowd, the strange sound their hoofs make on the asphalt, and the uncertain footing complicated by their iron-shod hooves striking the unfamiliar surface. For many years, Morris Ledbetter drove his 1946 red fire truck by, children hanging off its sides, and siren blaring. Bands play, and children wave from the wagons and trucks.

The replica of the Texas Central Railroad engine and cars "steams" down the street, its tractor chassis disguised by the skillful hands of its builder, Mr. Graham P. Crutchfield, and smoke billowing from the stack. In later years the squadron of Confederate Air Force planes from Abilene would stage a fly over. This is typically a grand parade, and spirits are high.

The parade follows the main street, which is also Highway 283. Beginning north of town past the old depot, now a museum where *Fandangle* scripts and other items are stored, the parade winds through the downtown area to circle the square where stands Albany's unique stone courthouse, the work of architect J. E. Flanders of Dallas begun in 1883. The crowd gathers on the ground around the courthouse by the gazebo to eat a catered barbecue dinner, prepared by Joe and Sharon Allen, specialists in such cooking. The menu is typical: sliced barbecued beef brisket, German sausage, pinto beans, coleslaw, sliced peaches, bread, and tea or coffee. Those early and lucky enough to find a place, sit at tables and chairs along the street to eat their meal. Others hunker down on the ground or around the gazebo. Conversation is the main order of the day. People who see each other daily or only once a year at the *Fandangle* or perhaps not in ten years or more choose this gathering to share news, pictures of grandchildren, and old memories of past performances witnessed or participated in. As the sun slants toward the west, the town empties of people as cars stream along the street leading to the parking lot of the Prairie Theatre on land once trodden hard by the hoofs of trail herds headed up the Western Cattle Trail and later by herds driven to the nearby railroad corral. Patrons navigate around the chartered buses and motor homes parked near the top of the hill. Shuttle trailers drawn by tractors scurry by carrying those who are unable or prefer not to walk up the gravel road toward the amphitheater. Friends greet each other with hugs and handshakes. The air, still usually in the ninety degree range, is nonetheless filled with expectation; and the audience eagerly anticipates the reenactment of the frontier settlement of the area.

THE HISTORY OF THE SHOW

How this show came to be created and produced initially and how it has continued to thrive over the years is a remarkable story in itself. In 1937, the superintendent of schools, C. B. Downing, called to his office one of the teachers, Miss Alice Reynolds, daughter of one of Watt's relatives. A graduate of Baylor University with a degree in music and with graduate work in music from Sul Ross State University, Miss Reynolds was a little puzzled by the invitation. And justifiably so, for the request was indeed a strange one: could she create and produce an outdoor drama with high school students as the cast members? She immediately admitted that she could not, but she recommended instead someone who could if persuaded. She then went to visit a good friend of hers, Robert Edward Nail, Jr., another offshoot of a locally prominent family, who had just moved back to his home town to see after his ill mother. He jumped at the chance. Shortly, Alice and Bobby, as they were affectionately known by scores of cast members, were off and running, and so was established the forerunner of the *Fort Griffin Fandangle,* a play originally entitled *Dr. Shackelford's Paradise,* done as the senior class play in the spring of 1938.

The creator left a rich legacy. Salutatorian of his high school class, Nail had attended Lawrenceville Preparatory School in New Jersey and Princeton University, where he studied drama and was named to Phi Beta Kappa. While at Princeton, he wrote *The Time of Their Lives,* regarded by many as the best play ever written by an undergraduate student at Princeton. Nail worked at theatres in New York as well as in Fort Worth, Abilene, and Dallas before returning in 1937 to Albany, where he taught English and drama temporarily at his old high school. He authored various one-act plays that proved popular, especially in the state's University Interscholastic League competition. In 1966, he was awarded an honorary Doctor of Humane Arts degree by Hardin-Simmons University in Abilene. Throughout the years, he received numerous other awards and was responsible for arranging the college education of numerous Albany youths and starting several on careers in technical theatre.

The performance of *Dr. Shackelford's Paradise* was so successful that by popular demand it was reworked to include townspeople in the cast and became known as the *Fort Griffin Fandangle,* an alliterative name that rolls euphoniously from the the tongue. Actually the last word is a variant use of the Spanish term *fandango,* a fast dance. A musical comedy from its inception, early versions of the show contained only traditional songs and dances, folk material at its best. Familiar tunes included "Whoopie Ti Yi

Yo, Git Along Little Dogies," which served as the basis for the "Huppi Hi" dance routine; "Those Texian Boys"; and "Eatin' Goober Peas." The theme song for the show, still used today, is a version of the tune to "Bury Me Not on the Lone Prairie," a song used as the theme in the early John Ford film *Stagecoach,* which led the way to stardom for actor John Wayne.

The show was so successful that Nail agreed to continue the production. But he laid down three rules from the beginning: first, anybody could be in it as long as the person had ties in Shackelford County; second, there would be no publicity, and the show would have to grow by word of mouth; and third, there would be no profanity. As the show was presented year after year by popular demand, participants began to write new songs. Now, most of the material is uniquely that of the *Fandangle,* although traditional songs are still sometimes included.

In his pamphlet entitled "*The Fandangle:* A People's Theater," Nail explains that the show evolved as a kind of "indigenous art" that "grows from the need of a community for dramatic expression" which "becomes a theatrical statement for the community's heritage and character." Of its development Nail said, "In the beginning it was meant to be a fairly serious spectacle, another pageant based on local history. It was, to use a cattle range expression, 'loose-herded' as it grew, and as often occurs in the creative process, it became not what was originally intended, but something much better." He explained this development by saying that "somehow the spirit of the men who made local history – their cocky individualism, their hard humor, their skeptical attitudes – kept intruding and turned bland pageantry into light-hearted satire." But history remained the basis of the drama.[6] Bob Green, local rancher, historian, and long-time narrator of the show, feels that participation of local youngsters in the show over the years has "painlessly indoctrinated several generations of children with the history of the area."[7]

Annual performances of the show have not been without interruption. Nail recounts in his pamphlet that by the late 1960s, thirty or so years after the first production, only sixteen shows had been produced.[8] When Nail entered the military service in 1942, the first hiatus occurred. During his military career, he was assigned to special duty writing plays and radio show scripts for the army. For this, he won the Legion of Merit medal. Following the war, he also created the *Nativity*, an annual Easter pageant, put on by the citizens of Albany. Occasionally, other years were skipped for various reasons. In 1953, the tenth performance was held. In 1960, an article in the *Breckenridge American* reported that the *Fandangle* would be revived after a three-year lapse, a hiatus probably triggered by the poor health and subsequent death, in 1958, of Mr. Crutchfield, a mainstay from early years of the production and the creator of several of the major props for the show.[9] Further, no show was done in 1969, the year following Nail's death on November 11, 1968. The shocked town mourned him as did anyone else who knew of him and his work. The board of directors rallied, however, and under the direction of Watt and others the show was revived and has not missed a scheduled production since that time. The drama has survived the loss of both major originators and original contributors, and Watt is one of the major reasons for that survival.

Nail saw the show as an activity unique to Albany and resisted efforts to "export" it to larger audiences. One conciliation was the formation of *Samplers*, shortened versions of the show that were transportable to other settings for rodeos and festivals, either indoors or out. These *Samplers* were generally produced early in the spring and really served as the core around which the larger show, usually two hours or more in length and involving about two hundred people, was built. *Samplers* have been given in many locations on the road, but the most significant ones were at the Lyndon Baines Johnson Ranch in 1967 and 1976 and in England in 1980. In addition, the *Fandangle* cast presented a show in Palo Duro Canyon and led the way, under Nail's leadership, to the massive professional production entitled *Texas* written by Paul Green and staged on a specially constructed stage in the canyon each summer. Although Nail was the catalyst who created and held the *Fandangle* together for many years, he did not work alone. Albany has more than its share of talented people who have contributed over the years.

Next to Nail, Alice Reynolds definitely made the greatest contribution. In addition to playing for many of the shows over the forty-three years in which she was active, she wrote many of the songs added in later years. Some she composed alone; on others she collaborated with James Ball, with Nail, and with both men. She designed many of the stage props as well, some of which are ingeniously constructed to allow movement on the stage. She made the drawings for and supervised the making of many of the elaborate costumes, though others certainly designed costumes as well. One of her major contributions is the design of the replica of the Texas Central Railroad engine and cars which Mr. Crutchfield built on a discarded tractor chassis. She also laid out the emblems of the fiddle and bow and the longhorn skull for the banners associated with the production, symbols which capture the spirit of the show itself. Miss Reynolds played for each performance without sheet music of any kind, using instead only a cue sheet. Once when she forgot the music and key of a particular song during a performance, the singer, Harold Law, started by himself perfectly in key and gave her the needed stimulus. In early performances she played only a piano with a microphone set beside it; later an organ became the source of the music. Although a longtime accompanist for the show, she has by no means played for all of them. For instance, Lloyd Letz played for some of the shows in the late 1930s and the 1940s. Alice died on May 20, 1984.

Others also loom large in *Fandangle* history. Reilly Nail, a nephew of the creator, started assisting with the show when he was still a Princeton undergraduate. A 1940 program contains names long associated with the show: costumes and scenery designed by John Nicholson; lighting by C. B. Downing, the high school superintendent whose idea the first production had been, and G. C. Burrage; and sound by John King. I. M. Chisum was stage manager, and Gaylena Bridges was in charge of costumes. Ollie Clarke and J. Carter King, Jr., handled publicity and business affairs. Narrators for that early show were Jean Anderson and LaVern Capps. Familiar names are found for those in the flag parade: Paul Armstrong, Jr., Harry Key, Mary Anna Green, Ray King, Jeanette Caldwell, Lynn Leech, H. C. Arendt, Doris Hill, and, of course, Watt Matthews. George Newcomb, longtime cowboy on Lambshead Ranch, was the old-timer in the 1940 production.

Through the years, some of these have continued their long association while newcomers have also become involved. Johnny Musselman also served as director for a time. Marge Bray, talented dancer and choreographer, who was in the first performance, is now the director of the show; Bill Overton, a college-trained drama director from Abilene, has contributed significantly;

James Ball has directed the show and has authored songs; Louann George assumed the duties of music director after Alice Reynolds retired in 1981. The list goes on and on. One good example of a longtime supporter is local rancher, Bill Green, who had been in all of the productions until 1990. His brother, Bob Green, and Marilyn Jacobs were narrators of the show for many years, but they have been replaced in recent years by Quay and Connie Parker.

The Prairie Theatre was not the original site used. The early shows were held on the Albany High School football field from 1938 to 1965. Mesquite limbs erected to form the backdrop for the open stage provided visual barriers behind which actors gathered for succeeding scenes and props were hidden. Then the one-acre outdoor theatre was built on a thirty-acre site northwest of Albany on land leased from Watt's family for a dollar a year. Built by the people of Albany under Nail's supervision, it is a unique and efficient theatre arranged to permit the diverse activities and sets used in the show. Not every stage will accommodate running horses, a host of dancers, a herd of longhorn steers, the Butterfield Stage, and the replica of the Texas Central Railroad train. The battery of spotlights and the flexibility and quality of the sound system far outstrip what one would expect of the resources of a town of 2,500 people. When one considers the support generated among Albany's business community, the reason for this quality becomes obvious.

The content of the show each year depends upon a general theme chosen by the directors and the talent available. Nail would often just pick out someone present at early rehearsals and assign a role. One of Nail's strengths was the ability to spot talent and tie it to some part of the show. Now Marge Bray is in charge of casting, in which she uses to her advantage knowing special and effective talent that has developed through performances over the years.

The technique of presenting the historical material in the *Fandangle* is a simple but effective one. The event or period is mentioned, often with factually correct dates and persons, and then the song or dramatic skit presents the human side of the event. This technique is especially noticeable in the Camp Cooper and Fort Griffin segments. Also a characteristic event such as drought is dramatized by the song such as "Windmill," the title and words of which evoke recollections of one of the emblems of the developing West. This large wheel-powered pump, first on wooden and later on metal towers, pumped water from beneath the earth and enabled ranchers to run livestock in fenced pastures away from natural water sources such as creeks and rivers. The earthen tank catching rain runoff is less often romanticized because it lacks the appeal of the windmill; besides, the tank became prominent only later in the development of the West. This focus on the romanticized version of events is often in defiance of reality, as the use of the windmill for the water source in the *Fandangle* indicates. In reality, little subsurface water suitable for drinking exists in the Albany area, and few windmills are therefore found. Instead of the subsurface water is the brine found in association with oil deposits in the area.

Favorite songs and scenes are repeated annually, but new songs and scenes are regularly added. The theme, however, is always the same—depicting the raucous, challenging life and developments characterizing settlement of the area. The details are simply variations on this theme of depicting the past. Even the format of the show makes a unique contribution, for in addition to listing the members of the cast and the scenes involved, the program, now in a newspaper tabloid format, is filled with information from old local newspapers—*Frontier Echo, Fort Griffin Echo, Albany Echo, Albany Star,* and *Albany News;* all early newspapers and, except for the *Star*, linked to each other through their historical development.[10] In the past, the program has differed in size from a pamphlet of six-by-nine inches in 1940 to a nine-by-twelve inch format in 1965.

The production schedule for a show usually begins in November and December when the directors begin talking about what to include. Then in January and February, casting and rehearsals begin as a *Sampler* is put together. The *Sampler* is always presented to members of the association at Reynolds Bend on the Clear Fork in May. Expansion of the show follows soon after, picking up speed as the school year ends and more actors from the schools become available. Nail always scheduled the show by figuring three minutes for one song and then timing the narration to be included to make up the desired length.

The old clock in the courthouse tower marks off each *Fandangle* performance, just as it has done the lives of Albany citizens over the more than one hundred years since the construction of the courthouse. The year's performance over, the crowd departs, people already promising to return next year to this most unusual event. As they leave, former cast members and family comment upon whose former costume was in what scene and was worn by whom, thus creating a sense of community with past performances. The *Fandangle* is so carefully woven into the fabric of life in Albany that to think of one is to automatically associate it with the other.

The production has received considerable attention in print in newspapers and magazines. An article in a 1965 issue of the Dallas *Times Herald* by A. C. Greene,[11] one of Texas' leading writers, was very complimentary; and Sam Pendergrast, one of Texas' most colorful news writers, authored two articles in the *Abilene Reporter News* in 1970, calling the *Fandangle* "pure magic."[12]

The myths, legends, and history of the frontier in this area are the gist of this production, and they are depicted in a nostalgic, romanticized fashion that eases the realization of the end that came to many institutions and individuals depicted in the story. It is, the narration says, "history not as told in books, but as remembered by the Old Timer," and, as Nail wanted, it is "people's theater." We remember the good times and learn in later years to laugh at the bad. That is human nature, and the good nature of the Griffinites depicted in this production makes the *Fandangle* what it is, a celebration of life in the Fort Griffin area of Texas. That emphasis suits the tastes of Watt Matthews, and he has given the *Fandangle* a level of support which has certainly helped make it what it is. By doing so, he may well have saved this monument in stories and songs for posterity, as he has done with other monuments in stone and wood.

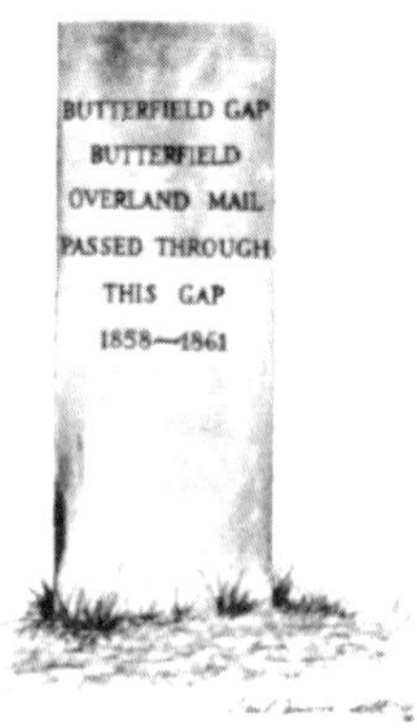

CHAPTER VII

ANOTHER SIDE OF A COW COUNTRY GENTLEMAN

The interests of Watt Matthews are as diverse as his accomplishments. In addition to his activities in ranching and historical preservation, he has long had other outlets for his energies. His excellent personal library housed at the headquarters complex of buildings and his activities in the cultural affairs of the community and of the world at large are two major areas of endeavor. He has, in addition, been instrumental in seeing two books important to the area into print. Although these efforts might seem atypical for one of Watt's background, this side of the rancher's image is not surprising when one considers the background and abilities of the man involved.

The most significant book with which he has been associated, of course, is *Interwoven*, his mother's chronicle of the settlement of the region. According to Carol Lackey, nine hundred copies of the first edition were privately printed in 1936 by the Anson Jones Press in Houston; 800 were in hardcover and 100 were in simulated leather. These latter volumes are now valued from $800 to $1200 each. One copy was bound especially for the author. Since interest in *Interwoven* remained strong after its initial publication, Watt determined to see the book published in a fine edition. At this point, he learned of Carl Hertzog, the famous El Paso printer, from historian Carl Coke Rister. "Hertzog and I became good friends," Watt says. "I became convinced that Hertzog was the one to put *Interwoven* into a fine edition, and I'm the one who put him onto Buck Schiwetz to do the illustrations. Buck had been to the ranch several times in those early days, when the Humble Oil Company had this ranch leased, and he had taken pictures of the structures. Some of these photographs came out in *The Humble Way,* a small magazine done by the Humble Oil Company.

Hertzog's talent produced in 1958 an excellent edition of 1500 copies in the Hertzog tradition with superb drawings by Schiwetz. Some of these volumes were bound in leather. In 1973, the University of Texas Press published a third edition and did a second printing in 1977, still carefully retaining the Hertzog look.[1] Texas A&M University Press reissued the book in 1982 in the same format.

In addition, the same press printed a companion volume, *Lambshead Before Interwoven,* in a very similar format. Written by Frances Mayhew Holden, this book is illustrated with drawings similar to those in *Interwoven* but done by Austin artist John Guerin. Mrs. Holden was assisted in her research by Watt, his brother, his sisters, and other members of the family. Copies of the original sets of drawings for both books now hang on the walls of the main ranch house at Lambshead. The two books, originally available in a boxed set, serve as a unique record of an historic time, place, and people.

Three other books reflect life on the ranch. *Summers at Lambshead* by Watt's nephew, John Burns, the son of Arthur P. and Susette Matthews Burns, was done by Burns himself. Another Hertzog piece is *Reminiscences* written in 1958 by Ethel Matthews Casey but not printed until 1977. This finely reproduced book contains a text and several pictures evoking the rich memories of the family and their relationship with the ranch. Watt, the ranch, the family, and the crew of cowboys have been depicted as well in the stark photographs made by Dallasite Laura Wilson and published along with a brief text in *Watkins Reynolds Matthews of Lambshead* in 1989. Other incidental maps and information sheets have been printed over the years in order to highlight some part of the history of the ranch.

It is *Interwoven*, however, which set the pattern for much of this printing, and Watt has definite memories about the creation of it. He recalls that once his mother started to write the book in 1934, she stayed busy at the task to the extent that she got cramps in her arms from her almost constant writing of the manuscript in longhand. Watt recalls that his father erected a metal bar outside the house so that when she walked around outside she could stretch her arms. In 1936 the manuscript was typed largely by Joe Blanton, a nephew of Watt, who also later edited *True Tales of the Frontier*, a fine edition of excerpts from *Interwoven*, published in 1961 to celebrate the centennial of Mrs. Matthew's birth. The original longhand manuscript of *Interwoven* no longer exists in a single collection because Mrs. Matthews presented sections of it to each of her children as part of their inheritance from her. Watt treasures the few pages that he received, for he holds a fond place in his heart for a mother who inspired him to pursue his own education and to be proud of his heritage, one well worth preserving.

Recently, when Watt received a questionnaire asking what he thought were the three best books on Texas, he showed an understandable personal bias when he listed *Interwoven* first. Many would agree with him. For the time in which it was done, it was a remarkable work in the tradition of Agnes Morley Cleveland's *No Life for a Lady*, Susan Shelby Magoffin's *Down the Santa Fe Trail*, Mary Kidder Rak's *A Cowman's Wife*, and more recently, a fictionalized account, *Filaree* by Marguerite Noble. Watt's efforts have kept *Interwoven* in print, and its acceptance continues to grow.

For the other two books on the list, Watt chose A. C. Greene's *A Personal Country* and John Graves' *Goodbye to a River*, both important works dealing with West Texas by two of the best contemporary writers of which Texans can boast and who are frequent visitors to Lambshead. Obviously, Watt has kept up with more than just ranching and the oil business.

Because of his personality and activities, Watt has cultivated friendships with numerous dignitaries, writers, and scholars. He became acquainted with Lyndon and Lady Bird Johnson, who visited Lambshead in 1967 and again in 1970. He accompanied the *Fandangle* cast on a trip to the Johnson Ranch in the Texas Hill Country for a *Sampler* during L.B.J.'s tenure as president, as well. He recounts that Lady Bird is "as fine a lady as I have ever known." In 1968, Governor and Mrs. John Connally also visited Albany for the *Fandangle* but did not visit the ranch. Famous

Country and Western entertainer Minnie Pearl came to the ranch at one time also, for her picture is carefully preserved in one of the photograph albums. On May 3, 1980, Watt hosted a party at Reynolds Bend honoring Dr. Lauro Cavazos, newly installed president of Texas Tech University in Lubbock, and members of the board of regents, only one more example of the kind of social role played by this man. Innumerable friends and acquaintances have enjoyed parties at the headquarters, at Reynolds Bend, or at one of the fishing camps located on the river. Watt has become familiar with other individuals important to the cattle business in the state. As a director for the state of Texas on the board of the National Cowboy Hall of Fame in Oklahoma City, he served with Robert S. Kleberg, Amon G. Carter, Jr., John Justin, and A. M. G. Swenson, quite an impressive list of ranching people that in other years included members of the Tandy family.

Other famous or important people are also in this circle of acquaintance. History interests Watt, so his association with historians is natural. One of the interviews for this project was interrupted by a phone call from J. Evetts Haley, highly respected historian of the region, who, I learned from the conversation, despite his advanced years, had just been bucked off a young horse he was breaking to ride. Watt's advice to Haley was "stay off those young horses." The historian was calling to see if a friend of his could visit the ranch. Watt's response was characteristic: "You have him call anytime after 4:30 in the morning. We'll be up and will tell him how to get here. He is certainly welcome to come." Dr. Rupert N. Richardson, one of the premier Texas historians and a man about ten years older than Watt, was a frequent visitor to the ranch until late in his life and is fondly remembered by Watt.

Carl Coke Rister, noted historian who ended his career at the University of Oklahoma, was a Sweetwater, Texas, native and was briefly a professor at Hardin-Simmons University. Their acquaintance began in an unusual way. In his classic work, *Robert E. Lee in Texas*, which deals in large part with Lee's activities at Camp Cooper, Rister comments on the drought of 1856. When he first met Rister, Watt remembers asking him what source he used for commenting on the drought. Rister cited Lee's correspondence now housed in the Library of Congress. To that comment, however, Watt responded, "To a man from Virginia, it may have looked droughty out here that year, but the real drought didn't come until two years later." Watt was not sure how Rister took this gentle criticism. Indication of just how Rister did take this comment is the dedication to Watt of a subsequent book by Rister, *Fort Griffin on the Texas Frontier,* which deals significantly with the Matthews and Reynolds families as well as with others in the region. This notice stemmed principally from this earlier experience and mutual respect of the two.

Watt has a circle of acquaintances developed through other activities as well. In addition to serving on numerous boards, he was made an honorary Texas Ranger of Company E in 1985. Sid Merchant, well-known Texas Ranger who served long years in West Texas, was instrumental in seeing that Watt received that honor. The award was presented at the Blake Field Ranch near Marathon in the Big Bend region of Texas at the annual gathering of Texas Rangers, not only the most prestigious law enforcement group in the state but one with Western mystique still clearly associated with it. Watt also boasts a certificate from the State of Texas naming him an honorary colonel of the governor's staff on February 22,1951. The document is signed by Pat Bullock as acting governor and John Ben Shepperd, Secretary of State. Although he appreciates these honors, the people he has met through them provide much of the pleasure he experienced through the recognition.

Watt continues to enjoy social gatherings and has seen to it that many friends and acquaintances have enjoyed the outings in which he has been involved. Although an exact chronicling of these events would prove too extensive to catalog here, some representative examples will indicate the scope of his activity. When the first edition of this volume was released in November of 1990, the autograph party hosted by Abilene Bookstore proved to be enormously successful. The crowd was large and sales were brisk, with well over a hundred volumes—more than a fourth of those printed—sold that day alone. People from near and far came and asked Watt to personalize their copy of the volume to a relative, to a friend, or to themselves. Watt complied with all requests in his distinctive handwriting style, even though the arthritis in his hand, which now precludes his signing any books at all, was beginning to bother him. He never complained.

Following the reception, Watt hosted at the Perini Ranch Restaurant in Buffalo Gap those who had come with him—Millie Diller, a long-time friend and companion who owns a neighboring ranch, and Bill Cauble and his wife Doris. Bill has lived and worked on Lambshead for almost a decade. Also present were Bill Curtis and Randy Armstrong of the HSU Press, and my wife and I. It was certainly a highlight for all of those involved as we enjoyed and the warmth of the fireplace on a cold, blustery day. Other activities have allowed happy outlets. Ranch rodeos have continued to play a role in the life of the ranch, and Watt is much in evidence at these. Area rodeos include Anson, Breckenridge, Albany, and especially the Western Heritage Classic in Abilene, where the ranch has for several years been in the chuck wagon cook off under the able direction of Bill Cauble. Bill's artistic skills often grace the invitations Watt sends out to various functions, especially to his *Fandangle* parties. Watt has been kind enough to allow the use of his wagon to serve supper (not aptly called "dinner" in this context) on the Saturday evening of the Western Heritage Classic in Abilene. This annual "gathering of the clan" brings together cowboys and friends for a period of visiting and eating barbecue.

Even Polo on the Prairie, a fundraiser for M. D. Anderson Medical Center for Cancer Research, is an event at which Watt will usually turn up and where he is thronged by well-wishers and friends. This event, held on the Musselman Ranch on the shores of Hubbard Creek Lake, features a polo match between a team representing the Musselman's Lazy Three Ranch and another, often the historic Spur Ranch located near the town of Spur. It is a festive event attended by several hundred people, many from Houston as well as from Dallas, Fort Worth, and Austin. Mary Ann McCloud of Eastland and others plan and host the event held in memory of Mary Anna Green Musselman, who succumbed to cancer several years ago.

Since 1990, guests have come to the ranch in numbers at least as large as in early years. Watt has welcomed the famous and the unknown. Chief of the famous was Nolan Ryan, the famous pitcher for the Texas Rangers, and several of his friends from the baseball franchise, who had their photographs made with Watt.

In 1991, Lambshead was one of five Texas ranches on a food tour by nineteen national writers on a Word from the Herd Tour sponsored by the Beef Industry Council of the Meat Board and the National Cattlemen's Association. The Texas Beef Industry hosted the tour. Writers with the group came from New York, California, Illinois, and Texas. Other ranches on the tour were the Double U, the Pitchfork, the Four Sixes, and the Brown Ranch.

Publications represented included the *New York Times, McCalls Magazine, Country Living Magazine, Longevity Magazine, American Health, U.S.A. Today, Ladies Circle, New Yorker Magazine,* and *Sunset.* Liz Weise, a nutrition news editor for CNN, was also on the tour.

In 1992, Russian poet Yevgeny Yevtushenko and his wife visited the ranch with a contingent from Hardin-Simmons University. I had arranged the visit but had not yet mastered the poet's name, so had not given it to ranch personnel at first. Bob Brittingham, Watt's nephew and quite an experienced man of the world and book afficinado, was waiting eagerly at the door of the cookhouse when the party arrived. When introduced to the famous poet, Bob exclaimed, "Yevtushenko! My God, uncle, this man is famous!" The poet enjoyed some of Watt's elixir and then toured the ranch. He was particularly impressed with the longhorn cattle and the wild game—deer, turkey, feral hogs, and quail. The evening meal and conversation were much enjoyed by all. The poet invited all to be at his poetry reading at Hardin-Simmons University the following evening and is already anticipating a return trip to Abilene and hopefully a return trip to Lambshead as well. This kind of listing would go on for a considerable time.

Not all of Watt's activities are on the ranch, by any means. Watt decided to make the trip for the seventieth reunion of his Princeton graduating class. The thought of going back at this point in his life had a certain appeal, and his health was still good, especially for a man his age. The mode of transportation differed noticeably from his first trip there seventy years earlier—he chartered a corporate jet airplane, and rather than go alone he took with him a party composed of Diana Nail, Sally Blanton Stopp, Mary Featherstone, Mary Ann Agee, Gerard Harrison, and Ted Blankenship, all friends or family members.

Festivities at Princeton were rewarding. Several of his classmates, most in wheelchairs and relying on oxygen bottles for support, were present. Watt made the rounds recalling old haunts and even rode in an open-top car in the parade honoring these graduates. His female guests were treated to a shopping trip in New York City. Gerard Harrison drove the limo but was cautioned by Watt to "put your money in your sock and be careful." One of the humorous developments was the questioning of members of the group about their attendance at the annual graduation ceremony of the university. Their negative reply was always followed by indicating their real purpose—the seventieth reunion of the Class of 1921. The trip was certainly a highlight for all involved.[2]

The considerable oil activity that has continued on the ranch brought about one very interesting set of circumstances. In 1991, a potential disaster was averted when on July 24, a well being drilled experienced a gas blowout at 5:00 A.M. This leak could easily have erupted into an inferno.[3] The pressure from the well was so great that it ruptured the capturing device designed to prevent this kind of accident. In fact, part of the pressure from this underground pocket escaped into a layer of gravel and found its way to an opening in the gravel bed on the Clear Fork near the Fort Griffin State Park. Gas bubbled up the water in the river and on a clear, windless July day collected under the bridge of the Clear Fork and forced closure of the bridge to traffic for a time because of the danger of an explosion.

One of the most interesting parts of this potential disaster was securing the services of Red Adair to direct the operation of capping off the well. The famous oil field fire fighter had to return from controlling fires in the Persian Gulf following Operation Desert Storm. At an enormous expense per day, the operation proceeded feverishly but carefully. Pumps and mud for the job were collected at a point almost a mile from the site of the well, while all the time a mist of natural gas hung over the rig. Carefully the necessary pipes and valves were fitted to the well, and finally on August 1, the moment of closing the valve was at hand. A worker clad in a protective suit and wearing an oxygen tank for breathing descended into the cloud to close that valve. He succeeded, but when the crisis was over the worker discovered that of the twenty or so minutes of supply of oxygen in his tank, he had consumed over half of it in the three or four minutes he had been in the gas cloud. Everyone supposed that he had been breathing quite rapidly.

In the tradition of generosity that he has evidenced many times over the years, on June 27, 1992, Watt agreed to the use of the William Reynolds Camp on the Clear Fork of the Brazos River to be the site for the fiftieth wedding anniversary party of Morris and Dorothy Ledbetter, longtime citizens of Albany and friends of Watt. On a hot but beautiful day, the honorees and guests were fed a meal of broiled steak with all the trimmings hosted by the Lawrence Clayton family. It was a gala affair in the beautiful setting on the river bank. Watt came and enjoyed the activities. It was on this same site in 1990 that the Claytons hosted with their chuckwagon the gifted and talented students from Haskell High School.

In 1992, Randy Armstrong and Bill Curtis from the HSU Press, along with my wife and me, realizing the significance of the moment, arranged through Bruce Cheeseman, archivist of King Ranch, Inc., a trip that would allow members of King Ranch management to visit Lambshead. It seemed appropriate that members of one famous ranching family visit the other. The King Ranch party consisted of Tio Kleberg; his wife Janelle; Roger Jarvis, C.E.O. of King Ranch, Inc.; and Leonard Stiles, retired cattle manager of the Santa Gertrudis Division of the famous South Texas ranch, along with Mr. Cheeseman.

The group brought gifts for Watt. One was an unusual vest, which he enjoyed wearing immensely. In fact, it became one of his favorite items of apparel. They also brought him a copy of the Saddle Blanket Edition of Tom Lea's famous two-volume history of the King Ranch. Watt graciously accepted it and then showed the group his earlier copy signed by the author, Carl Hertzog, and Robert Kleberg, Jr.

Watt reciprocated their generosity and took them on a tour of the ranch, an experience the visitors found fascinating. Of particular interest to the group was the historic preservation possible in the relatively dry climate of West Texas, but impossible in the humid heat of South Texas. The HSU Press later printed fifty copies of a broadside to celebrate this historic event. This piece, none of which were sold to the public, is a treasured item for those who have one. Each sports the signatures of Tio Kleberg and Watt. It was the reenactment of a custom once associated with royalty, where members of one family pay homage to the other one.

In 1993, a contingent from Corinth, Greece, in a sister-city program with Abilene, was invited to the ranch to tour the buildings and to see the wild game, as well as to eat lunch at the cookhouse. A traditional meal of chicken fried steak, mashed potatoes, and all the trimmings prepared by Joy Bundy, the current cook, was a highlight for the visitors who managed hats and boots as well as denim jeans for the trip. Their first view of a real live cowboy was of Wes Farley, a young cowboy on the ranch who looked every bit the part, because he is. It was indeed a delightful part of the Greeks' visit to Texas.

Over the years Watt has continued to host the annual party for members of the Fort Griffin Fandangle Association on a Friday night in April in Reynolds Bend. The event is eagerly anticipated by all who have been to one of these. The meal, in later years catered by restauranteur and superb chuckwagon cook Tom Perini of Buffalo Gap, is always followed by a performance of a *Sampler* of the show.

On the following Saturday night Watt hosts a private party for cast members and guests. The event is certainly appreciated by all who are invited to enjoy Watt's hospitality. On Saturday, May 1, 1993, the private party was interrupted by a violent thunderstorm that sent spectators fleeing to their cars after only fifteen minutes of the show. All of the guests escaped the rain and mud except for one party from Fort Worth, who took a wrong turn and got stuck in mud. Ranch personnel had to help them out after being notified in a circuitous way. The man in the car called friends in Fort Worth on his car phone, and the friends called the ranch to sound the alarm for the stranded party. All ended well, and the visitors had an extra to go along with their *Fandangle* party. It was an interesting evening for all. The ranchers in the area appreciated the rain, although the storm blew limbs off trees and rattled roof tops.

One of the interesting developments in the history of Fort Griffin itself is the forming of Company F, 4th U.S. Cavalry (Memorial) to honor the unit led by Col. Ranald MacKenzie, the famous Indian fighter, who inflicted a sound defeat on the Comanches in Palo Duro Canyon. Watt has been made an honorary member of the unit. In 1992, a group rode from Fort Griffin to the site of Camp Cooper in an reenactment of a ride that occurred originally on September 25, 1871, when Capt. Wirt Davis, commander of F Company of the 4th U.S. Cavalry, left Fort Griffin with more than a hundred and twenty troopers to meet MacKenzie at the Camp Cooper site. From there, the unit left on the first campaign of the famous Red River Indian Wars.

In addition, a marker approved by the Texas State Historical Commission indicating the route called the MacKenzie Trail was erected on April 25, 1993, on the Howsley Ranch adjacent to Lambshead and under the management of Dick Hash. Also, interest has been directed at a campsite used by MacKenzie at the base of the hill upon which Fort Griffin stood. A dig has recently been held there and turned up various artifacts.

Another interesting social gathering occurred in May of 1993 when Watt and his sister, Lucile Brittingham, were honored at the unveiling of an historical marker at Bud Matthews, Texas, a set of railroad corrals erected in the early 1900s southwest of Albany. Ella Matthews Conrad, wife of the man who operated a store in Fort Griffin and later in Albany, and a sister of J. A. Matthews, owned and managed this ranch after her husband's early and unfortunate death. Ella and J.A., whose nickname was Bud, combined efforts on the pens that saw thousands of cattle, even one small herd of buffalo, loaded onto rail cars headed to Fort Worth or Kansas City. The pens were restored in the 1920s by Grady Smith, and again in the early 1990s by the Driscoll family, descended from G. R. Davis, who bought the ranch in 1920.

The event honored these early day ranching greats. The Driscoll family located an authentic cattle car near Amarillo and had it restored by Edward Spurgin of Albany and moved to the site, an especially appropriate act since few of the cars have survived from the heyday of railroad shipping and the tracks were removed in 1967. Spokesmen for the day's activities were Edgar and Ted Paup, sons of Mary Frances "Chan" Driscoll, owner of the property. A reception and catered barbecue meal were hosted by the family at the headquarters of Chimney Creek Ranch, where Robert Waller, and his son, Robert C. Waller, run a quality cattle operation.[4]

Watt recalls visiting his Aunt Ella at the ranch headquarters when he was a boy. To make the trip, the family rode in a wagon following the ruts left by the Butterfield Overland Mail coaches, for the route on its way from St. Louis to San Francisco ran through Lambshead and on to Smith's Station located on the Chimney Creek Ranch property. The Driscoll family also had a marker commemorating the Butterfield route moved to the site of the Bud Matthews Switch.

Media attention has continued for Watt. In July of 1992, a crew from the Washington, D.C., offices of the Austrian National Television Network led by Franz Koessler came to Lambshead and the Nail Ranch to film a documentary on contemporary cowboy life. After filming an early morning roundup and cow working on the Nail on July 24, Franz and his crew, including camera man David Robinson and sound man Johnny Ludwig as well as Franz's friend Monika, traveled to Lambshead to interview Watt.

Koessler considered the interview with Watt extremely informative, and made Watt's remarks a central part of the documentary he edited for presentation in Austria in early 1993.

Watt has not had pain-free later years, but he has endured his inevitable aging with characteristic verve. Of his longevity he has often said, "If I die today, I am way ahead of the game." Of his aches and pains, he has commented that men who die young do not have the opportunity to experience them. Two major problems have plagued him. Never a heavy eater, Watt began losing weight in the early 1990s, and he did not have much weight to lose. Alarmed, he went to Fort Worth to undergo medical tests and discovered that his problem was tied to an ulcerated valve that was causing his stomach to discharge food before it was digested. He had other problems associated with the aging process but dedicated himself to regaining his health. Responding to his medication, he began to gain weight very gradually and soon resumed his active role as master of Lambshead.

One humorous incident related to his recovery involved a late afternoon "toddy" session in which all the guests were having some refreshment, except for Watt who was drinking the vitamin supplement prescribed by his physician. As he fretted about missing his evening ration of Crown Royal, the physician, Dr. Clark Gregg, one of Watt's nephews, who was present, responded, "Uncle Watt, you can have some in your supplement." Watt responded with characteristic vigor, "Well, put it in," he said. Everyone thought his recovery perked up from that moment on.

Watt also began to walk to regain his strength. All of his efforts were aided by the careful supervision of Mary Featherstone, a nurse from Albany who moved her motor home adjacent to Watt's room in the bunkhouse to see after him. She has been his constant companion and social secretary since her arrival at the ranch. For years the nurse of Dr. T. M. Howell, respected Albany physician, Mary had been retired for several years and had also suffered the death of her husband. Because of these circumstances, she was able to devote her time and energy to nursing Watt, a job she has done particularly well. She has also become his social secretary and keeps a very busy calendar of activities on the ranch. A good example of the pace of these activities occurred in 1993 following a late night ride back from the Western Heritage Classic in Abilene and an early rising the next morning to attend another social event. Mary was heard to say, "I guess I'm going to have to start taking his medicine to keep up with him."

Another problem concerned cataracts that begin to dim Watt's vision. Undaunted by the effects of aging eyes, Watt arranged with an Abilene physician to remove the problem. Surgery on the first eye went well, and later the second operation came off without any unusual complications. Not many weeks after the second surgery, I was lucky enough to be driving a small group on a tour of the ranch in Watt's vehicle. He was seated beside me in the passenger seat. As we stopped to examine the Hoover house, he looked down at the small calendar from a local oil field supply house. The calendar was overall no larger than three by four inches, but the portion recording each month was about one inch high and less than three inches wide. Without delay he read, without his glasses, the days and the phases of the moon recorded on that tiny piece of paper. He asked me if he was correct, and I had to put my glasses on to see if he had read it properly.

Certainly not all of Watt's time has been spent in social outings and the like. For the most part, he has been seriously involved in ranching and in the promotion and breeding of Hereford cattle. Largely because of his commitment to Hereford cattle, Watt has also been very active in the American Hereford Association. In 1950, the Grand Champion steer at the Fort Worth Stock Yards came from the Lambshead herd. Displayed on the wall at the main ranch house are several trays awarded to Watt by the Fort Worth Fat Stock Show and Livestock Exposition for having championship animals in various categories in 1957, 1960, 1961, 1962, 1963, and 1964. In 1975 the Hereford show at the Southwestern Livestock Exposition in Fort Worth was dedicated to him for his "many contributions to the Hereford industry in Texas and the United States." It is an honor that Watt certainly deserves, as is the honorary life membership awarded the same year by the Texas Hereford Association. He served as a director of the American Hereford Association from 1960–1966.

Watt has supported cattle-related projects as well. He holds an honorary Lone Star Farmers Degree from the Texas Association of the Future Farmers of America, awarded on July 16, 1959. He was honored by the Cattlemen's Roundup for Crippled Children through the West Texas Rehabilitation Center in 1975 and has a certificate dated June 27, 1974, from the Council on International Relations for service to the Fort Griffin Fandangle Association. Numerous other plaques for similar awards dot the walls.

Watt is proud of the ranch and feels it is a particularly attractive one. I have met no one who disagrees with that belief. The promoters of Marlboro cigarettes found the rusticity of Lambshead Ranch particularly suitable for their advertising needs, and Watt allowed them access to the resources there. In earlier days when cigarette commercials could be aired on television, the company made arrangements to come to Lambshead to shoot some footage. Company executives were originally attracted to the area because they wanted to capture on film the longhorn herd at nearby Fort Griffin. When the crew was at the 6666 Ranch near Guthrie, Texas, the manager, Jack Pumphreys, recommended that they get in touch with Watt because he would likely be able to help them make whatever arrangements were necessary.

When the advertising executives came to consult with Watt over the matter, they liked the layout of the ranch and began looking for ways to use the setting in their advertisements. They soon inquired as to whether the ranch owned a chuckwagon. Watt acknowledged that they did, and he recommended his cook, John White, for the cook's job in the advertisement. The staff wanted an older man for the job, and the director asked Watt if he himself would be interested in playing that role. He declined. Watt told them that he would help them make a quality commercial but that he would not appear in a cigarette advertisement. He said, "I have given the boys so much trouble about smoking here on the ranch that I couldn't see helping promote the use of tobacco." However, Roy Matthews agreed to play the role of cook in the commercial.

The Lambshead chuckwagon, which is used only on special occasions, was rolled out to an appropriate spot at the mouth of Paint Creek, and the camera started rolling at daylight. Several cowboys played parts in the commercial, but Roy Matthews had a particularly good time in his role as chuckwagon cook. When the camera crew prepared to depart, the promoters gave Watt $500 to parcel out among the men who had participated with indication as to how much was to go to each person. By the time Watt got to the end of the crew, only $10 was left for Matthews. He was undaunted from the small amount because he had enjoyed the activity so much; however, the advertisement caught the attention of the public, and Roy Matthews' royalties from that appearance eventually amounted to almost $10,000. Evan Easter, nicknamed Egg, a cowboy on the ranch at that time, also caught the fancy of the advertising executives, and he was photographed on several occasions. His net return from this activity was about $3,500. Watt has been more than willing to share attention with those around him.

Even after the television commercials were discontinued because of a federal ban of such advertising, the company continued to return to Lambshead to shoot some of the still shots that have come to be the mainstay of their commercial campaigns. Pictures made on Lambshead have appeared in Marlboro advertisements on the back covers of *Time, U.S. News and World Report,*and *Southern Living.* The positive response to these realistic advertisements has been strong. For example, even though *Western Horseman,* one of the most prestigious magazines dealing with western life, refuses to carry advertisements for cigarettes, the magazine gave Marlboro a commendation for showing western life in a dramatic way in those commercials, many of which were filmed on Lambshead Ranch. The promoters formed a practice of bringing their own models, and on several occasions a male model with heavy red mustache, seen in many commercials in the mid 1980s, appeared on Lambshead to be the subject of these still shots. It is small wonder the Marlboro Company was enchanted with Lambshead Ranch.

The ranch is in an almost idealized setting because of the rolling wooded hills and rocky creek beside the headquarters. Benny Peacock, a cowboy who rode on this ranch as well as other large ranches in Texas, said, "It's the prettiest ranch I ever worked on."[5] The cluster of structures there is appropriately outfitted for this setting. The buildings are arranged in a rough square, with well-kept barns and corrals to the west of the main buildings. On the north is the red brick bunkhouse and a long low building with screened porches. In the center is a cookhouse constructed of native stone. Watt's picket-style house stands to the west of the cookhouse. On the south is the original ranch house. Other structures to serve housing needs for some of the other families are arranged around this square. An underground storm cellar is in the southern part of the courtyard, and the concrete top of the cellar serves as a sitting area in the cool of the evening. A sundial and large rain gauge are on the western edge.

The focal point of the open space is a large piece of bronze statuary by Terry Kelsey, a talented western artist. The piece depicts a herd of longhorns going north on the trail. Among the cattle is an authentically outfitted cowboy sitting on a high-backed, slick-fork saddle appropriate for the period. The horse bears the AV brand of Lambshead Ranch, and each of the steers carries a brand identifying an early owner of longhorns purchased and placed on the Wichita Mountain Wildlife Reserve in Oklahoma to preserve the breed from extinction. This piece is a small version of the heroic-size piece that stands at the future home of the Texas Longhorn Breeders Association near Billy Bob's Texas in Fort Worth in the old Stockyard District.

How the piece came to be at Lambshead is an interesting story. Kelsey is a neighbor and friend of John Brittingham, Watt's nephew, who manages the Park Springs Ranch in New Mexico. Brittingham, who earlier lived near Kelsey in Colorado, became friends with the artist and sold him a piece of land on the ranch upon which the artist built a home. After Watt became interested in Kelsey's work, he and Brittingham bought the piece of statuary now located at Lambshead. Kelsey did seven more of these remarkable pieces in table-size models. Several other pieces of Kelsey's work are located in the First National Bank in Albany, a facility outfitted with a pioneer-era facade and decor. Watt is a longtime director of the bank.

The original ranch house is, as one observer noted, "a treasure trove of Western memorabilia."[6] Old-style saddles are found in various places around the house, and authentic Indian wool rugs cover the floors. Several lever action Winchester rifles, representative of the frontier period, hang on the walls; an old Colt six-shooter lies on top of a low cabinet in the kitchen. An impressive library of books, old and new, occupies one room; and numerous scrapbooks and photograph albums that chronicle the story of the ranch and of the West are housed in this structure. One of these books is a study of the works of Carl Hertzog, the famous El Paso printer. It is a fine boxed edition of Al Lowman's *Remembering Carl Hertzog,* which takes its place alongside other books and numerous magazines containing articles about life on the ranch. Among these are *American West* by Richard Avedon, *A Salute from Above* by T. R. Fehrenbach, *Texas* by Robert Reynolds, and *Jericho: The South Beheld* by Herbert Shuptrine and James Dickey.

Numerous paintings, prints, and photographs decorate the walls. A set of Charles Russell prints shows the sequential development of the West from the beginning to the coming of the railroad, an important frontier development. These are entitled *Before This He Had Only Wolves Broke To Pack or Drag a Travois, To Fort Union to Trade, The Pony Express, The Stage Station, Freighting from Fort Benton, First Fire Canoe Nears Fort Union,* and *The Iron Horse Comes to the Upper Missouri.* The last picture in the grouping is not by Russell but is instead by Tom Lea, a well-known Texas artist, and is entitled *The Airplane.* This print shows a cowboy standing beside his saddled horse and watching a World War II-vintage bomber flying over. Thus is represented the settlement of the frontier from the days before Indians had horses to the coming of modern aviation. One of the most remarkable of the many photographs of Matthews catches him beside the picket-style house on the ranch. This enlarged photograph, a startling and revealing character study, hung in the Old Jail Art Museum in Albany for a time.

Other publications containing pictures of Watt and his cowboys can be found in this collection. A framed copy of a longhand version of Berta Hart Nance's poem "Cattle" occupies a place on the kitchen wall, a reminder of one of Albany's more illustrious former residents and of the opening of the *Fandangle,* where the poem is recited each year. Also in the house is an upright concert cabinet grand piano. Lettering indicates that it was made specifically for J. A. Matthews of Albany, Texas, in 1885, by Boardman and Bray of Albany, New York. On the piano is an eighteen-inch-high bronze statue of a cowboy in full gear, and on the wall behind it hangs a buffalo robe, a fitting meld of culture. Over the mantle in the living room is a drawing of a herd of longhorn cattle by Frank Oliver. The predominant decor is definitely that of the West.

Hanging on the wall of the main ranch house is an artistically printed document entitled "Wattisms,"expressions for which Watt has become well-known: "Are you well?" "How you come on?" "Plumb handy," "Bravo," "Golee," "Weyno," "Plumb alright," "Why don't you sell the Charolais now? The price is higher than a tree," "He is not even a notch on a stick," "They're about to talk

Paul Cameron Smith '88

'em up a storm," "She's a little swole," "Shellin the woods," "The fat boys fell out," "She had a rough go," "Give it a lick," and "Pay no tensch." Decorating the document are the AV and gourd brands as well as the imprint of Carl Hertzog, who prepared the document.

One of the many art works hanging on the walls of the house is a drawing of two cowboys dragging a split beef carcass over the leading edge of a raging grass fire, a grim reminder of a situation that still threatens ranchers today. Also displayed on the wall is a certificate from St. Andrew's School in Fort Worth, Texas. The heading proclaims it to be a "Certificate of Proficiency." The document certifies that "Master Watt Matthews has attained an average of 85% in the following studies: arithmetic, grammar, reading, spelling, geography, U.S. history, Texas history, and Bible history during the session of 1909–1910." The document is dated June 1, 1910, and is signed by Elizabeth Whitaker and Cora Bowie Fitzhugh, Principal. Also noticeable is the Josiah Wheat Award of Merit signed by Gov. Preston Smith. In a small frame is a letter on grayish paper complete with the envelope. It is addressed to Col. H. H. Craig, Chief of Ordnance, Washington City, D.C., from Camp Cooper, Texas, 1 February, 1857. The text is as follows: "Colonel, Have the honor to transmit herewith quarterly return of ordnance and ordnance stores for the second quarter of 1837. Respectfully, your servant Robert E. Lee, Bvt. Col., 2nd Cavalry." One of the documents is an honorary life membership from the Fort Griffin Fandangle Association, dated 1968. On the wall are prints of paintings by Tom Luvell and Melvin Warren, and a beautiful original by Mondel Rodgers, a Sweetwater artist, of a branding crew showing Watt wielding the branding iron on a calf held on a working table. The Ranger Award from the Texas and Southwestern Stock Raisers Association is a pearl-handled Colt .45 pistol, preserved under a glass dome. A bust of Watt—about 10 inches tall—entitled "The Old Breed" by Jimmy Cox, area bronze sculptor, stands in the sitting room of this ranch house, a "treasure trove" indeed.

The cookhouse on Lambshead, built years ago to replace the original one, has been renovated and expanded, and a truly modern kitchen and storeroom provided. In 1993, additional space was constructed north of the cookhouse in a new stone structure that replaced a small storage building. Provided are efficiency sleeping quarters for the cook to accommodate early morning meals and office space for Watt and Carroll Martin. This rustic cookhouse, which had served as Watt's office, is open to travelers just as it was in the previous century. The hospitality is indicated not by an enamel coffee pot hanging from an iron rod over a camp fire but by an electric coffeemaker that stands ready to yield a cup of steaming brew.

Heating in the winter is provided for the cookhouse by large fireplaces on the north and south ends of this building, and many a steak has been broiled on coals dragged from the fireplace onto the cement floor in front of the north fireplace. The andirons carry the AV brand of Lambshead Ranch. Along the east wall in the winter months stand large stacks of seasoned mesquite wood ready for use. This wood adds its aroma to that of the food and aromatic smoke. Along the eaves inside the cook shack hang rifles on antique racks made from horseshoes and in other Western designs. Replicas of various pieces of statuary by Remington and Russell stand on the south mantle.

Visible near by is a framed photograph of Charles Goodnight with W. D. Reynolds, one of Watt's uncles. The men were good friends, and the picture was made when Goodnight stopped by to accompany Reynolds to a convention of old-time trail drivers in San Antonio sometime before 1929. The two had been well acquainted for many years. One of their unique experiences is reflected in a letter written by Reynolds to Harry E. Crain in 1925, Reynolds recounts how he and Goodnight brought the exhumed remains of Oliver Loving, Goodnight's partner in establishing the Goodnight-Loving Cattle Trail, from Fort Sumner in New Mexico Territory to Weatherford, Texas, for permanent burial. The pioneer cattleman died as the result of wounds received in a fight with Indians along the Pecos River near the Guadalupe Mountains. He was taken to Fort Sumner by a group of Mexicans traveling in ox carts, but soon succumbed to blood poisoning.[7] There seems little doubt that Larry McMurtry fictionalized this actual event in his popular novel *Lonesome Dove*.

One of the most significant and unique recent additions to the complex of buildings at the headquarters is a picket-style house made of old railroad cross ties set vertically and chinked with mortar. Watt has prepared this facility in case he becomes incapacitated, and he seriously calls it his "nursing home," since the facility can accommodate a wheelchair. The building has a rustic appearance, and Watt recalls that a visiting architect inquired about the age of the structure. The visitor assumed that it was quite old and was amazed to learn that construction had been completed only days before. Inside the building is a large living/bedroom combination with a fireplace in the north end. A sawed mesquite timber varnished to a rich amber serves as the mantle. On it stands a statue of a Hereford bull, almost twelve inches tall. The inscription reads: "For Watt Matthews, Champion Hereford Breeder, Exhibitor, Rancher, in Recognition of Outstanding Service to the Livestock Industry of America at the Watt Matthews ROM Hereford Show, Fort Worth, 1975, Presented by the American Hereford Association." Other pieces of statuary stand around the room. A bronze longhorn skull, its horns stretching to impressive lengths, by Covelle, stands on a small table; and a bronze sculpture of a wild tom turkey, its beard clearly evident, stands on a stone base on the mantle. Two twin beds sit on the cement slab floor covered by Indian rugs. The bathroom occupies the south end. On the west wall, framed under glass, is a quilt made by Sallie Reynolds Matthews for one of Watt's sisters. Standing along the east wall is a large rolltop desk. On the top of this desk is a piece of pewter sculpture entitled *The Holdup of the Overland Stage* by Gordon Phillips. The base carries the notation "Created especially for Watt R. Matthews." Also located near by is the Western Heritage Wrangler Award from the Cowboy Hall of Fame and Heritage Center in Oklahoma City for Watt's efforts to preserve history through the *Fort Griffin Fandangle*.

Above and on both sides of the desk is an extensive collection of books. To the left are works by John Updike, Wallace Stegner, Carl Sandburg, Nathaniel Hawthorne, James Fenimore Cooper, Ernest Hemingway, John Steinbeck, Pearl S. Buck, and others. The most notable part of the collection, however, is the impressive array of Texana and Western Americana. Histories include Tom Lea's *King Ranch*, J. W. Williams' *The Big Ranch Country*, Carl Coke Rister's *Fort Griffin on the Texas Frontier*, and Rupert N. Richardson's *This I Remember*. Others are Edgar Rye's *The Quirt and the Spur*, Dorothy Emerson's *Among the Mescalero Apaches*, John C. Dawson's *High Plains Yesterday*, John G. Bourke's *On the Border with Crook*, and David Dary's *Cowboy Culture*. An especially fine volume bound in leather with the hair out and laced on the edges with a leather string is C. L. Douglas' *Cattle Kings of Texas*. David

Murrah's *Pitchfork Land and Cattle Country* takes its deserving place on the shelf as well. A particularly favorite writer of Watt's is Dr. Curry Holden, husband of Frances Mayhew Holden, author of *Lambshead Before Interwoven.* Once at McMurry College and later at Texas Tech University, Dr. Holden has produced works including *Hill of the Rooster, The Espuela Land and Cattle Company, Rollie Burns, The Spur Ranch, Alkali Trails,* and *A Ranching Saga: The Lives of William Electious Halsell and Ewing Halsell.* The rare Roscoe P. and Margaret B. Conkling's *The Butterfield Overland Mail,* the definitive history of the stageline that ran through the region and crossed Lambshead Ranch, is also present.

Other categories are included. Works of fiction are not numerous but are of high quality: A. B. Guthrie's *The Way West,* Conrad Richter's *The Town,* A. C. Greene's *Highland Park Woman,* J. Frank Dobie's *A Vaquero of the Brush Country,* and Bill Brett's *The Stolen Steers. Stand Proud,* a novel by San Angelo novelist Elmer Kelton, is also appropriately in the collection. This volume tells the story of a rancher whose life parallels to a remarkable extent the early ranching experiences of the Reynolds and Matthews families. The novel opens at Fort Davis in Stephens County on the Clear Fork where the Reynolds family "forted up" during the Civil War. Kelton acknowledges that he used the Newcomb diary, part of which is included in *Interwoven,* as the basis for the beginning action and setting of the novel.[8]

Particularly noticeable in Watt's collection are biographies of figures significant to the region: Lon Tinkle's *An American Original: A Biography of J. Frank Dobie,* C. L. Sonnichsen's *Cowboys and Cattle Kings,* John Graves' *Hard Scrabble,* Charles Siringo's *A Texas Cowboy,* Barney Nelson's *The Last Campfire,* C. M. "Buck" Newsome's *Shod with Iron,* Leon C. Metz's *John Selman, Texas Gunfighter,* Willie Newberry Lewis' *Tapadero: A Study of Her Husband, An Early Day Cattle King,* Chris Emmett's *Shanghai Pierce,* Mary Whatley Clarke's *John Chilsholm,* and in a finely bound volume *A Stoveup Cowboy's Stories* by John Emmett McCauley. The introduction for this book was written by John A. Lomax, famous folk song collector; the afterword was penned by Carl Hertzog. The book is graced with drawings by Tom Lea. Other books in the collection are *Will Rogers* by Richard M. Ketchum and *George Littefield* by J. Evetts Haley.

Several art books grace the shelves as well. They range from a volume of Norman Rockwell's paintings to *Taming the West: A*

Photographic Perspective, edited by David H. Phillips, and Charles M. Russell's *Trails Plowed Under*. Also included are *The Frederick Remington Book* and *The Charles M. Russell Book*, volumes capturing the work of two of the world's most famous early-day western artists. Buck Schiwetz's *Texas* and the *Schiwetz Legacy* are found alongside *Landscapes of Texas* and *Texas in Bloom*. The books on Watt's shelves range from older volumes to the contemporary works of John Erickson, from Perryton, Texas. Housed in this rustic structure is a fine tribute to the life and times of the pioneers and particularly to the life of a modern-day frontiersman such as Watt, a representative of that ongoing tradition. This collection will match or exceed that of most private collectors.

It is not surprising that a man like Watt Mathews is well acquainted with the literature of the West and with a broad range of people from a wide area of the world and that the beautiful country of Lambshead draws people to the ranch, for he has made a unique contribution in enhancing the unique natural beauty of Lambshead and developing the ranch headquarters complex of buildings into what they are today. Watt's friendly hospitality and firm support of worthy causes has in turn brought him recognition and honor, a fitting tribute to one deserving the attention and fame.

CHAPTER VIII

PLANNING FOR THE FUTURE

Even in his mid-nineties, Watt plans no retirement from his work. If he becomes incapacitated, he has prepared a place to accommodate himself on the ranch. Further, he has strong feelings about staying at Lambshead as long as possible because of the way he feels about his home and the ranch that has been that for him these many years. When he dies, he will be buried in the family cemetery along the Clear Fork, though he does not plan for that to occur any time soon. Since most of his brothers and sisters have lived well into their nineties, he is planning to follow in their footsteps. His sister Lucile is now well past one hundred.

In an interview with David McCullough on a Public Broadcasting System program entitled *Home – Where the Heart Is,* Watt reiterated his intentions to hold the place together for his descendents. He fully intends to pass this ranch on in better shape than when he got it to those generations. He sees Lambshead Ranch as his heritage, a responsibility he does not take lightly. The PBS program on which he expressed these views was one of national significance because the two other people included on the program were then-San Antonio Mayor Henry Cisneros, whom President Bill Clinton appointed as Secretary of Housing and Urban

Affairs; and a former First Lady and a long-time friend of Watt, Lady Bird Johnson, company in which Watt Matthews is as much at home as he is in the branding pen. Watt will turn over a rich legacy to his inheritors, one better in several ways because he took charge of it. But the legacy is there because he cared and used his resources wisely. The deep tracks of Watt Matthews are present for all to see and to admire.

Watt is a man who loves music, people, parties, good food, and good livestock. He is at home on the ranch and in the finest restaurants and ballrooms in the country. Pat Lidia Jones, in an editorial in the *Albany News* to commemorate Watt's eighty-eighth birthday, stated, "Watt has been Albany's most eligible bachelor for over seventy years. He was when he was eighteen and he still is. Pursued by many women over the years and captured by none, he remains one of the most desirable escorts on our not inconsiderable social scene." She goes on to credit him with being "a great dancer, a good conversationalist, an attentive date, and unfailingly kind. He is a gentleman, and he has impeccable manners. Who could ask for anything more?" Her praise for Watt is high indeed. She said, "He has probably kissed and been kissed by more women than I have ever known—that number exceeded only by the number of men who have shaken hands with him.

"For Watt is not just a lady's man. He is also a man's man. If he shows his age at all, perhaps it is because he still believes a man's word is his bond and a handshake over a business deal is as binding as the signature on a legal agreement. His is."[1] She goes on to note that anyone who asks about Albany always inquires about Watt, for the two are inseparably linked.

Part of his hospitality is visible at the Fort Worth Livestock Exposition, where he has regularly hosted a party for friends and associates at this famous outing that captures the spirit of the Southwest. He is interested in art as well, a fact attested by the numerous art works around the ranch headquarters and in the various buildings. But his range of interest exceeds his personal holdings, and he has interest in the arts elsewhere. He says of the Old Jail Art Foundation Museum in Albany, for example, "There's not a town in America that wouldn't be proud of that museum. Albany is a small town, but it has big ways." He is proud of that. Watt has come a long way in bringing a degree of luxury to the Clear Fork. His father's idea of luxury, sleeping between clean sheets and drinking ice water, has been well surpassed by modern life on the ranch. But even in this, Watt has maintained a simple existence. He does not consider himself a pioneer. "A true pioneer moved out here and decided to stay," he says. "We were born out here and did not know anything any better." But a Princeton man has seen a different world to know whether it is better for him or not. And Watt decided to stay on the Clear Fork.

About the future of ranching and the beef industry, Watt is optimistic. He says that the raising of beef is a basic industry and that though the bloom of the often romanticized business may be over, people will continue to eat beef whether the ranches on which that beef is raised are corporate or private. He admits that keeping this new type of ranch going will not be any easier than running one of the older ones because, he says, "It is the standard rule that about every twenty-five or thirty years about half the farmers and ranchers go broke." He also recalls that his father said: "It is darn easy to get land poor." His family was land poor until about 1902 before any stability came to this region, and Watt intends to see that the stability remains.

The story of Watt Matthews covers most of a century, one that has seen rapid development in every area of life. But amid this flurry of change he has not only remained stable but is as well a stabilizing influence on his family and friends. As he presides over a restful period of conviviality in the spacious rooms of the original ranch house in the evenings while someone cooks steaks over mesquite coals in the cookhouse, one can sense the strength of the man and the institution of ranching in West Texas. Here is western life the way it should be, not glamorized on the silver screen, but lived in the quiet vastness of the Clear Fork country, one of the finest cattle-producing areas in the world and the home of one of the world's greatest cattlemen – Watkins Reynolds Matthews. Bravo, Watt, and Weyno!

PHOTOGRAPHS

(All photographs courtesy of Sonja Irwin Clayton.)

A modern roundup on Lambshead, May 1985.

Watt applies the ranch brand to a Lambshead calf, May 1985.

A cattle baron surveys his herd, May 1985.

The flock of sheep kept at the headquarters to keep grass and weeds down as well as to continue the heritage of some sheep on the ranch.

The cookhouse from the west, before the extension of 1993 to the north.

The Terry Kelsey sculpture in the courtyard of the headquarters.

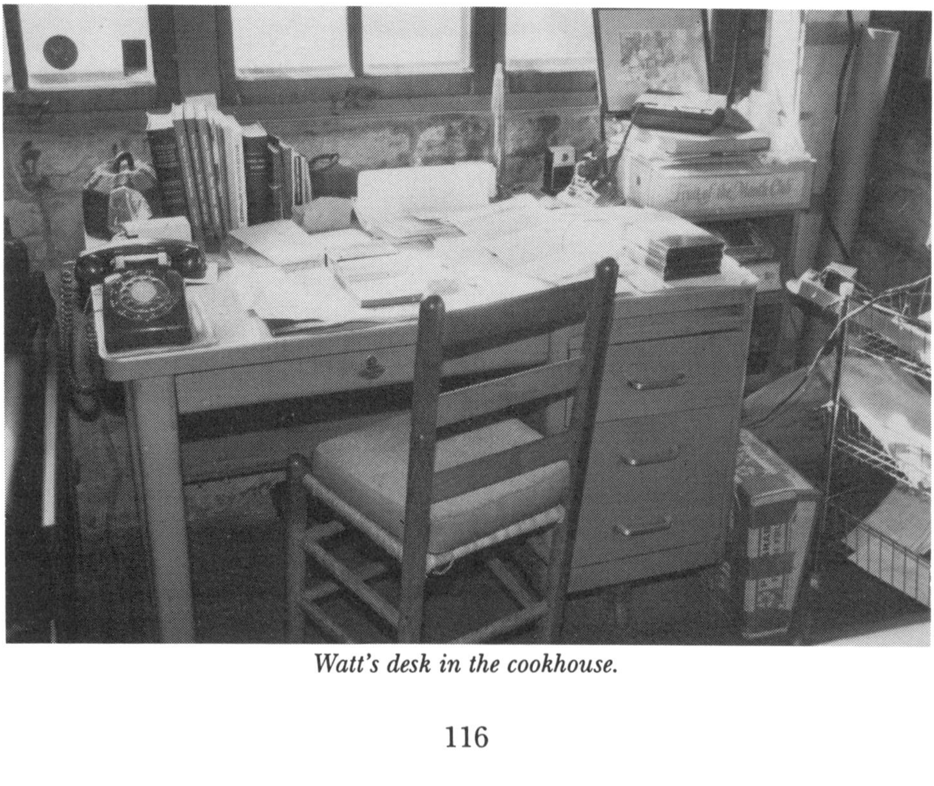

Watt's desk in the cookhouse.

View from the west of the picket house at the headquarters.

Watt's personal library in the picket house at the headquarters.

North fireplace in the cookhouse.

The schoolhouse in Reynolds Bend.

Marker for the Butterfield Stage route.

The bunkhouse at the headquarters. Watt lived in the room on the right for most of his life.

Front view, facing east, of the main house at the headquarters.

A front view of the dugout.

View from the south of the Matthews Memorial Presbyterian Church in Albany.

Watt and Mary Featherstone in foreground at a western art show and dinner at Hardin-Simmons University in 1988.

John Vandiver at Bennett Camp on the river.

Bill Cauble at Bennett Camp on the river.

Stone Ranch from the south.

Buffalo lounge around Stone Ranch.

Detail of the stacked stone on the corral at Stone Ranch.

Aubrey Lange's helicopter on the ground during a roundup.

Terry Moberley ropes out a mount in the main corral at the headquarters.

Watt sitting on the base of the Terry Kelsey sculpture.

Watt reading his mail on the couch in the cookhouse. Note his desk in the background.

Watt enjoys the company of the author (left) and George Peacock (right) at George's party on Nail Ranch, March 20, 1992.

The stagecoach, a prop for the Fandangle, *makes its way down Albany's main street during the parade. Terry Moberley and Dick Hash sit on the driver's box.*

Mr. Crutchfield's version of the Texas Central Railroad train for the Fandangle.

Watt awaiting his turn on the program at the graduation ceremony at which he received his honorary doctorate at Hardin-Simmons University.

Watt and the author in a familiar pose, recording the history of the man and the area.

NOTES

PREFACE

1. Newsletter, Center for Texas Studies, North Texas State University. Ed. A. C. Greene, Elizabeth Gunter, James W. Lee. No. 3, Summer, 1987. p. 3.

CHAPTER I

1. Sallie Reynolds Matthews, *Interwoven,* 4th ed. (College Station: Texas A&M University Press, 1982), p. 193.
2. Phin W. Reynolds, "Phin W. Reynolds to J. R. Webb," Unpublished manuscript, Hardin-Simmons University Archive, p. 1.
3. *Interwoven,* p. 37.
4. Carl Coke Rister, *Robert E. Lee in Texas.* (Norman: University of Oklahoma Press, 1946), p. 83.
5. Carl Coke Rister, *Fort Griffin on the Texas Frontier* (Norman: University of Oklahoma Press, 1956), p. 63.
6. Phin W. Reynolds, "Recollections and Experiences of the Frontier Life of Phin W. Reynolds." As told to J. R. Webb, Unpublished manuscript, Hardin-Simmons University Archive, pp. 38–41.
7. See *Interwoven,* pp. 177–182.
8. Undated document in Barker Center, University of Texas at Austin.
9. Mary Whatley Clarke, "Uncle Joe McComb," *Cattleman Magazine.* June, 1974, pp. 64, 66–67.

10. Preston Lewis, "Lottie Deno, Gambler," *Old West.* September, 1987, pp. 16–21. See also Charles Robinson, III, *The Frontier World of Fort Griffin.* Spokane, Washington: The Arthur H. Clark Company, 1992, for a detailed treatment of the town.

11. Rister, *Fort Griffin,* pp. 156–157.

12. John Chadbourne Irwin, "The Frontier Life of John Chadbourne Irwin." As told to J. R. Webb. Unpublished manuscript, Hardin-Simmons University Archive, p. 4.

13. Lawrence Clayton and Morris Ledbetter, "John Calvin Ledbetter/S. W. Wesley: One Indian Captivity Story or Two?" West *Texas Historical Association Yearbook.* 60 (1984), 83–91.

14. Edgar Rye, *The Quirt and the Spur.* (Austin: Steck-Vaughn, 1967). See also Charles E. Linck, Jr., *Edgar Rye.* (Commerce: East Texas State University, 1972).

15. Bill Whitaker, "Calf-Branding Is No Act for Watt Matthews," *Abilene Reporter-News,* June 19, 1987, p. 3C.

16. Ruby McGill Dodge, *Reynolds Presbyterian Academy and College.* (Belton: Peter Hansborough Bell Press, 1960).

CHAPTER III

1. Undated manuscript in Barker Center, University of Texas at Austin.

2. Undated manuscript in Barker Center, University of Texas at Austin.

3. See Phin W. Reynolds, "Recollections and Experiences of the Frontier Life of Phin W. Reynolds." As told to J. R. Webb, Unpublished manuscript, Hardin-Simmons University Archive, pp. 26–27.

4. Ethel Matthews Casey, *Reminiscences.* (N.P., 1978), pp. 1–2.

5. Interview of Jack Pate by Lawrence Clayton, Albany, Texas, June 20, 1985.

6. Audrey Parker Brooks, "County Calf Lands in North Carolina," *Albany News,* Thursday, July 16, 1987, p. 6B.

7. Undated newspaper clipping in Matthews family scrapbooks.

8. Lawrence Clayton, "The Ice Man Comes to Fort Griffin: An Oral History From the 1930s." *West Texas Historical Association Yearbook.* 61 (1985), 141–146.

CHAPTER V

1. "Portrait of a Frontier Family," *House Beautiful Magazine.* January 1985, pp. 72–77, 115.

2. Francis B. Heitman, *Historical Register and Dictionary of the U.S. Army.* Washington, D.C.: U.S. Government Printing Office, 1903, p. 130.

3. The information on Givens was supplied by B. J. Fisher from the National Archives in Washington, D.C. Fisher was a long-time

archivist at this location and a serious student of the war records housed there. His assistance is gratefully acknowledged.

4. Fisher, Interview by Lawrence Clayton, November 9, 1988.

5. Fisher, Unpublished manuscript prepared from documents in the National Archives, p. 2.

6. Xerox copy of text in my possession furnished by Mr. Fisher from The National Archives.

7. A typescript of a report to the Adjutant General in Washington. November 17, 1854. Furnished by Mr. Fisher.

8. Francis Mayhugh Holden, *Lambshead Before Interwoven* (College Station, TX: Texas A&M University Press, 1982), p. 39.

9. General Order No. 2. War Department. Adjutant General's Office. Washington, D.C., January 15, 1856.

10. Regimental Returns. Second U.S. Dragoons. Copy available in Hardin-Simmons University Library.

11. Letter from Givens to Colonel C. A. May, December 26, 1985.

12. Special Order No. 2, May 15, 1856.

13. Heitman, p. 130.

14. Regimental Returns. Second U.S. Dragoons.

15. Letter from Dr. L. T. Heard, Galveston, Texas. November 24, 1857.

16. Letter from Bvt. Major General D. E. Twiggs from Colonel S. Cooper.

17. Holden, p. 87.

18. Heitman, p. 130.

19. Don H. Biggers, *Shackelford County Sketches* (Albany: *Albany News* Office, 1908), n.p.

20. Holden, p. 37.

21. Sallie Reynolds Matthews, *Interwoven* 4th ed. (College Station: Texas A&M University Press, 1982), p. 142.

22. "Matthews to Receive KTB Award," *Albany News*, June 25, 1987, p. 3A.

23. "Four West Texas Cowmen Honored by Wichita Falls Ranch Roundup," *Livestock Weekly*. August 20, 1987, p. 6.

24. Interview of J. C. Irwin, Jr., by Lawrence Clayton, Abilene, Texas, July 20, 1987.

25. Interview of John Vandiver by Lawrence Clayton, Albany, Texas, April 20, 1984.

CHAPTER VI

Much of the information for this chapter comes from an interview with Alice Reynolds in Albany, Texas, on May 12,1984, just a week before her death, and from examination of *Fandangle* programs in the Matthews collection. Another version of the origin of the show is that members of the senior class asked Nail to help them with a project to raise money for a class trip.

1. Berta Hunt Nance, "Cattle," *Fort Griffin Fandangle* Program. 1985, p. 8.

2. See Preston Lewis, "Lottie Deno, Gambler," *Old West*. September, 1987, pp. 16–21.

3. Robert Nail and James Ball, "The Town of Fort Griffin," in *Fort Griffin Fandangle Favorite Songs*. (N.P., 1966), p. 10.

4. J. Frank Dobie, *The Longhorns*. (Boston: Little, Brown, and Company, 1941), p. 35.

5. Robert Nail, "Prairie Land," in *Fort Griffin Fandangle Favorite Songs*, p. 1.

6. Robert Nail, "*The Fandangle*: A People's Theater." (Albany, *Fort Griffin Fandangle* Association, 1970), p. 6.

7. Bob Green, informal interview by Lawrence Clayton, December, 1991.

8. Nail, "*The Fandangle*," p. 5.

9. Undated news article in Matthews family scrapbook.

10. See Charles E. Linck, Jr., *Edgar Rye*. (Commerce: East Texas State University, 1972).

11. Undated news article in Matthews family scrapbook.

12. Undated news article in Matthews family scrapbook.

CHAPTER VII

1. Carol Lackey, "Family Book Relates Area History," *Albany News, Fandangle* Souvenir Section, 1987, pp. 3–5A.

2. Interview of Diana Nail by Lawrence Clayton, Albany, Texas, June 1, 1993.

3. Interview of Mary Featherstone, Lambshead Ranch, June 3, 1993.

4. See Lawrence Clayton, "Chimney Creek Ranch." Abilene: Privately Printed, 1993.

5. Interview of Benny Peacock by Lawrence Clayton, Albany, Texas, July 10, 1986.

6. Interview of William Curtis by Lawrence Clayton, Abilene, Texas, January 22, 1987.

7. Harry E. Crain, ed., *Letters From Old Friends and Members of the Wyoming Stock Growers Association*. (Cheyenne, WY: S. A. Bristol and Company, 1923), p. 54.

8. Interview of Elmer Kelton by Lawrence Clayton, Abilene, Texas, October 10, 1985.

CHAPTER VIII

1. Pat Lidia Jones, "Ponderings by Pat." *Albany News*, Thursday, February 5, 1987, p. 4A.

BIBLIOGRAPHY

INTERVIEWS

Much of the information in this text came from several interviews with Watkins Reynolds Matthews between 1985–1993. In addition, Mr. Matthews read the manuscript in July of 1987 and corrected errors in facts and dates. Material for the Second Edition was added after June 1, 1993.

Other interviews were conducted with Terry Moberley, foreman of Lambshead Ranch, on the conduct of the cattleworking; John Vandiver on game conservation; Richard Ballard, Billy Gilbreath, and Carrol Martin on everyday life on the ranch; George and Sue Peacock on "neighboring" with Watt; and Louann George on the *Fandangle* material.

Special help was provided by Alice Reynolds, whose knowledge of the *Fort Griffin Fandangle* was invaluable. The interview with her was conducted on May 12, 1984, just a week before her death. I am in particularly indebted to my father-in-law, J. C. Irwin, Jr., for many anecdotes concerning life in the area from the early decades of the century to the present.

Other interviews are indicated in the notes where appropriate. Special thanks go to Dr. David Murrah, director of the Southwest Collection at Texas Tech University, for wise counsel on research materials, especially for telling me about the photograph albums and scrapbooks housed at Lambshead Ranch; these materials have been microfilmed and are housed in the Southwest Collection along with other items related to the ranch.

SELECTED BIBLIOGRAPHY OF PRINTED MATERIALS

"Albany's Watt Matthews on 'CBS This Morning' Today." *Abilene Reporter-News*, November 5, 1992, p. 3B.

Baggett, Donnis. "A Good Guy Named Watt." *Dallas Morning News*, May 12, 1991, n.p.

Barranger, Donna. "Old Stone Ranch Restored." *Fandangle* Souvenir Section, *Albany News*, 1984, pp. 3-5A.

——. "Matthews Receiving Award." *Albany News*, February 14, 1985, p. 1A.

Biffle, Kent. "The Tale of a Boy and His Ranch." *Dallas Morning News*, October 15, 1989, pp. 43A, 46A.

Blanton, Joe, ed. *True Tales of the Frontier*. Albany, Texas: Venture Press, 1961.

Brooks, Audrey Parker. "County Calf Lands in North Carolina." *Albany News*, Thursday, July 16, 1987, p. 6B.

Burns, John. *Summers at Lambshead*. N.P., 1977.

Casey, Ethel Matthews. *Reminiscences*. N.P., 1978.

Clarke, Mary Whatley. "Uncle Joe McComb." *Cattleman Magazine*, June, 1974, pp. 64, 66–67.

——. "Lambshead." *Cattleman Magazine*, 55, no. 8 (January 1969), 44.

Clayton, Lawrence. "Chimney Creek Ranch: An Historical Account and Personal View of a Shackelford County Ranching Heritage." Privately Printed, 1993.

——. "The Ice Man Comes to Fort Griffin: An Oral History of the 1930's." *West Texas Historical Association Yearbook*. 61 (1985), 141–146.

——. "Bill 'Tige' Avery." *Cowboys Who Rode Proudly*. Ed. J. Evetts Haley, Jr. Midland: Haley Museum, 1992, pp. 7–16.

——. "Watt Matthews: Historical Preserver." *Heritage Magazine*, 4, no. 3 (1986), 30–31.

—— and Morris Ledbetter. "John Calvin Ledbetter/S.W. Wesley: One Indian Captivity Story or Two?" *West Texas Historical Association Yearbook*. 60 (1984), 83–91.

Crain, Harry E., ed. *Letters From Old Friends and Members of the Wyoming Stock Growers Association*. Cheyenne: S. A. Bristol Company, 1923.

Dobie, J. Frank. *The Longhorns*. Boston: Little, Brown, and Company, 1941.

Dodge, Ruby McGill. *Reynolds Presbyterian Academy and College*. Belton: Peter Hansborough Bell Press, 1960.

"Four West Texas Cowmen Honored by Wichita Falls Ranch Roundup." *Livestock Weekly*, August 20, 1987, p. 6.

Green, James Robert. "A Brief History of the Stone Ranch." N.P., 1984.

Greene, A. C., Elizabeth Gunter, and James W. Lee, eds. Newsletter,

The Center for Texas Studies, North Texas State University. No. 3. Summer 1987.

"Guardians of the Land." *Albany News*, May 17, 1990, p. 2.

"HSU Dean Hosts Austrian Film Crew." *Abilene Reporter-News*, July 29, 1992, p. 3A.

"HSU Dean To Host Gifted Students Today." *Abilene Reporter-News*, August 3, 1992, p. 3A.

Harris, Rebecca. "HSU Salutes Cowboy Spirit." *Abilene Reporter-News*, February 28, 1991, pp. 1B, 3B.

Holden, Frances Mayhugh. *Lambshead Before Interwoven: A Texas Range Chronicle, 1848–1878.* College Station: Texas A&M University Press, 1982.

Irwin, John Chadbourne. "Frontier Life of John Chadbourne Irwin." As told to J. R. Webb. Unpublished manuscript, Hardin-Simmons University Archive.

Jones, Pat Lidia. "Ponderings by Pat." *Albany News*, Thursday, February 5, 1987, p. 4A.

Lackey, Carol. "Family Book Relates Area History." *Fandangle* Souvenir Section, *Albany News*, 1987, pp. 3–5A.

"Lambshead Included on Food Tour." *Albany News*, May 30, 1991, p. 8.

Lawrence, Larry. "Matthews Is Subject of Two Books." *Abilene Reporter-News*, December 7, 1989, p. 15A.

Lenamon, James. "Watt Matthews and the Texas Tradition." *Persimmon Hill*, *4*, no. 4 (1974), 16–25.

Lewis, Preston. "Lottie Deno, Gambler." *Old West*, (September 1987), pp. 16–21.

Linck, Charles E., Jr. *Edgar Rye, North Central Texas Cartoonist and Journalist.* Commerce: East Texas State University, 1972.

"Living History Activities Planned at State Park." *Albany News*, April 22, 1993, p. 1.

Lockridge, Mary. "Galloping into the Past." *Fort Worth Star-Telegram*, April 14, 1992, pp. 17A, 20A.

Lucas, Donnie A. "Matthews Receives Presentation Award." *Albany News*, Thursday, May 10, 1984, p. 1A.

——. "Watt Matthews Given 2nd Award." *Albany News*, Thursday, May 31, 1984, p. 1A.

Lucas, Melinda L. "Memorial Regiment Re-enacts Past." *Fandangle Souvenir Section, Albany News*, 1993, pp. 6A–9A.

Maddox, Cynthia."Watt Matthews: Rancher." *Texas Highways*, 38, no. 4 (April 1991), 12.

"Matthews Receives Special Award." *Albany News*, August 13, 1992, pp. 1, 10.

Matthews, Sallie Reynolds. *Interwoven: A Pioneer Chronicle.* 4th ed. College Station: Texas A&M University Press, 1982.

"Matthews To Receive KTB Award." *Albany News*, Thursday, June 25, 1987, p. 3A.
McCoy, Dorothy Abbott. *Texas Ranchmen: Twenty Texans Who Help Build Today's Cattle Industry*. Austin: Eakin Press, 1987.
Nail, Robert E. *The Fandangle: A People's Theater*. Albany: Fort Griffin Fandangle Association, 1970.
Nance, Berta Hart. "Cattle." in *Fort Griffin Fandangle* Program. 1985, p. 8.
"New Biography Available." *Albany News*, December 7, 1989, p. 3A.
"Patel Chimney Service Takes on Matthews' Ranch." *Abilene Reporter-News*, May 4, 1993, p. 2C.
"Portrait of a Frontier Family." *House Beautiful*. January 1985, pp. 72–77, 115.
"Public Invited on Historical Ride." *Albany News*, March 26, 1992, p. 4B.
"Rancher Nominated to Hall of Honor." *Albany News*, April 2, 1992, p. 2A.
"Rancher Preserves History." *Fandangle* Souvenir Section, *Albany News*. 1984, pp. 5–7B.
Reed, Jerry Daniel. "HSU To Give Honorary Degrees to Two Big Country Pioneers." *Abilene Reporter-News*, May 5, 1989, p. 5A.
Reynolds, Alice, James Ball, and Robert Nail. *Fort Griffin Fandangle Favorite Songs*. N.P., 1966.
Reynolds, Phin W. "Phin W. Reynolds to J. R. Webb." Unpublished manuscript, Hardin-Simmons University Archive.
——. "Recollections and Experiences of the Frontier Life of Phin W. Reynolds." As told to J. R. Webb. Unpublished manuscript, Hardin-Simmons University Archive.
"Rider Training Session Set February 22." *Albany News*, February 20, 1992, p. 7A.
Rister, Carl Coke. *Fort Griffin on the Texas Frontier*. Norman: University of Oklahoma Press, 1956.
——. *Robert E. Lee in Texas*. Norman: University of Oklahoma Press, 1956.
Robinson, Charles, III. *The Frontier World of Fort Griffin*. Spokane, Washington: The Arthur H. Clark Company, 1992.
"Round Ups: Notice to Cattle Men [*sic*]." *Fort Griffin Echo*, August 23, 1879. (From a typescript in Barker Center, University of Texas at Austin).
Rye, Edgar. *The Quirt and the Spur: Vanishing Shadows on the Texas Frontier*. 1880, rpt. Austin: Steck-Vaughn, 1967.
Shockley, Marion Daniel. "The Fort Griffin Fandangle: A Historical Study of an Outdoor Drama." M.A. Thesis, Sul Ross State University, 1966.
Smith, J. T. "Area Ranching Legend Chosen Man of the Year." *Abilene Reporter-News*, September 21, 1992, pp. 1A, 6A.

———. "Conservationists Cited by Lower Clear Fork Soil and Water District." *Abilene Reporter-News*, December 12, 1989, p. 7A.

———. "Doctor Cites 'Cholestrolphobia.'" *Abilene Reporter-News*, June 2, 1991, p. 4C.

———. "Farm Periodical Takes a Look at Area Rancher Watt Matthews." *Abilene Reporter-News*, March 14, 1993, p. 8E.

———. "New York Writers Visit Area To Learn about Beef *and* Cowboys." *Abilene Reporter-News*, May 25, 1992, pp. 3A, 10A.

———. "Switch Dedication Honors Albany Ranching Family." *Abilene Reporter-News*, May 19, 1992, p. 8C.

———. "Western Heritage Events Give Time To Look Back on History." *Abilene Reporter-News*, May 5, 1993, p. 8C.

"250 Ride Old Trail to Camp Cooper." *Albany News*, April 16, 1992, p. 5B.

"Watt on Tube." *Abilene Reporter-News*, November 7, 1992, p. 4A.

Whitaker, Bill. "Calf-Branding Is No Act for Watt Matthews." *Abilene Reporter-News*, Friday, June 19, 1987, pp. 1, 3C.

———. "Poet Devours Slice of Texas." *Abilene Reporter-News*, November 17, 1992, pp. 1A, 8A.

———. "Photo Exhibit Reveals Life in Albany." *Abilene Reporter-News*. Sunday, December 2, 1984, p. 5E.

———. "Watt Matthews: Rancher Proves To Be Spirit of Lambshead."*Abilene Reporter-News*, April 18, 1991, pp. 1A, 6A.

———. "Old Rancher Likes Guests Dropping By." *Abilene Reporter-News*, May 4, 1993, pp. 1A, 10A.

———. "'Old Watt' Continues to Welcome Guests." *Abilene Reporter-News*, May 25, 1992, pp. 1A, 3A.

———. "Trailing Watt Had HSU Professor Back in Saddle." *Abilene Reporter-News*, November 27, 1989, pp. 1A, 8A.

Wilson, Laura. *Watt Matthews of Lambshead*. Austin: Texas State Historical Association, 1989; Excerpts reprinted in *Dallas Life Magazine*, November 12, 1989, pp. 12–16, 18–21.; and *Spirit*, June, 1990, pp. 91–94.

Wolfshol, Karl."Profile: Clinging to the Past for Today." *Progressive Farmer*, 108, no. 3 (March, 1993), 16.

Index